Gary L. Herstein
Whitehead and the Measurement Problem of Cosmology

PROCESS THOUGHT

Edited by

Nicholas Rescher • Johanna Seibt • Michel Weber

Volume 5

Gary L. Herstein

Whitehead and the Measurement Problem of Cosmology

ontos
verlag

Frankfurt | Paris | Ebikon | Lancaster | New Brunswick

Bibliographic information published by Die Deutsche Bibliothek
Die Deutsche Bibliothek lists this publication in the Deutsche Nationalbibliographie;
detailed bibliographic data is available in the Internet at http://dnb.ddb.de

North and South America by
Transaction Books
Rutgers University
Piscataway, NJ 08854-8042
trans@transactionpub.com

United Kingdom, Ire Iceland, Turkey, Malta, Portugal by
Gazelle Books Services Limited
White Cross Mills
Hightown
LANCASTER, LA1 4XS
sales@gazellebooks.co.uk

P.O. Box 15 41, D-63133 Heusenstamm
www.ontosverlag.com

ISBN 3-937202-95-1

2006

Printed on acid-free paper
ISO-Norm 970-6
FSC-certified (Forest Stewardship Council)
This hardcover binding meets the International Library standard

Printed in Germany
by buch bücher **dd ag**

Table of Contents:

Preface

This work originated as my dissertation, but it must be said that the research began long before I ever made the decision to complete my Ph.D. in philosophy. My interest in Whitehead first emerged over twenty-five years ago, as an undergraduate at Occidental college, but my interest in science and mathematics goes back to my earliest childhood memories. Yet despite this personal background, my introduction to Whitehead's criticisms of orthodox general relativity was something that only occurred a few years ago, once I had already begun my Ph.D. work. More importantly for me was the discovery that these criticisms have never been adequately addressed by the scientific community and, as I will argue in the following, remain essentially valid. This was a rather astonishing discovery for me, for like many people with a strong (if somewhat informal) background in physics, I had been under the impression that general relativity was basically "done," and that the only real problem that remained to be addressed was the reconciliation of gravity and macro-physics with quantum theory. Investigating the extent of this sanguine error has led to the following book.

There are a few people whose inspiration and help I would particularly like to mention at this point. Foremost among these is Dr. Randall Auxier, whose many years of guidance with the intricacies of Whitehead's thought, as well as general friendship, made this work possible. In addition, the editors at Ontos Verlag – Johanna Seibt, Rafael Hüntelmann, and Michel Weber, have been absolutely invaluable with their patience, interest and expertise.

Finally, I would like to dedicate this book to my Father, without whose faith and support I would never have been able to engage in this type of research.

Introduction

The notion of "any" frees us from individual character: but there is no entity which is merely "any." Thus, when algebra is applied, factors beyond algebraic thought are relevant to the total situation.

Alfred North Whitehead, "Mathematics and the Good."

It is the central thesis of the following work that there is a fundamental problem at the heart of general relativity that relates to the very possibility of generating meaningful cosmological measurements; measurements, that is to say, that are fully interpretable and logically coherent within the framework of general relativity and its presupposed philosophy of nature. This claim is easily misunderstood, as are the purposes of the following argument. The above claim does not relate (directly, at least) to the empirical results when numerical calculations based upon the general theory of relativity are compared with various observations. Rather, it relates to the underlying basis by which it is assumed that those numerical calculations and refined observations are interpreted and related the one to the other. Moreover, while the following argument will draw heavily upon various works by Alfred North Whitehead, the following is not intended as an exercise in Whitehead scholarship. While some care will be taken to situate the following within Whitehead's thought and justify associating this project with process thought in general, our task here is to use Whitehead's ideas as a set of tools with which we might analyze a problem and sketch a generic outline of that problem's solution.

This introduction is intended to preview the journey we are about to take, providing us with, if not a map, at least a schematic list of the major turns we will be making. One of the first moves to be understood on this journey has to do with the notion (mentioned above) relating to the *nature* of the criticisms that are going to be advanced here. Our concern is not with the empirical adequacy *per se* of general relativity, but rather with its logical coherence. This latter concept needs to be explained. It will often be stated in this inquiry that the measurement problem we are investigating is a *logical* problem. But the use of the term "logical" here must not be supposed to be relating exclusively to the formal manipulations of symbols. This is the modern notion of logic as the science – or at least study – of deductive relations, validity, and the like. For us, however, the notion of "logic" that we will be explicitly calling upon hearkens back to a much earlier time when logic was the study of valid methods of *inquiry*.

The historical roots of this position are most clearly present in the logical works of Aristotle, while the most prominent contemporary advocates are John Dewey and Jaakko Hintikka. This way of looking at logic can be a source of some confusion, particularly as some of these issues have come to be grouped under the heading of "methodology." However, this latter term does not do the work we need done here – it is at once too broad and too narrow. Meanwhile, it is necessary to have some categorizing label to mark off inquiry, and inquiry *into* inquiry, from mere mathematical deduction. The term "logic" has both historical and systematic reasons for taking on this role. We will explore some of these reasons in greater detail at the end of chapter one, and I beg the readers patience with this usage until then.

In any event, the claim that the theory of measurement in general relativity suffers from a logical problem does not purport to assert that there is some type of common or widespread deductive failure with respect to the formal consequences of the mathematical theory, or that the numbers that some scientists have produced (and which many others have repeatedly checked) are somehow wrong. Rather, what is being claimed is that, from the philosophical perspective of the theory of inquiry, general relativity cannot legitimize the results it produces. Given what the theory assumes regarding the "nature of Nature," the contents of human experience, and the validity of generalizing various relational structures under the constraints of space and time that general relativity posits, the theory of measurement in general relativity is incoherent. The nature of this incoherence, some salient aspects of its history, and Whitehead's responses to it, are the central themes of this book.

Now, in the course of this investigation we will have cause to examine a few of the historically important mathematical formulae relating to general relativity. However, while we will examine some of the formal features of contemporary cosmology, at no time will we be *doing* physics. An analogy that might serve here is this: our work is akin to that of a literary critic, but we are not composing a novel. Nothing that occurs in these pages will ever aspire to the level of a mathematical proof, even in that comparatively more forgiving sense of "mathematical proof" which governs much of the physical sciences. We are outsiders taking a peek inside the box of contemporary physics and cosmology, trying to understand what we are witnessing and even offering some educated commentaries on what we are observing, but not directly tinkering with those inner workings. It is for others to do such work.

So, while we will be moving in and out of the formalisms of physical cosmology, we will not allow ourselves to be trapped within them. Indeed, we can quite appropriately trust the mathematicians and physicists to keep the formal deductive aspects of the mathematics in good order. While it is certainly the case that mathematical errors are made by even the best individuals and teams among these groups, these are precisely the sorts of things that the physics and mathematics communities are ideally suited to discover and eventually weed out. Typographical mistakes, lapses of authors' attention, insufficient background in specialized fields of mathematical argumentation on the parts of copyists and proof readers, are all just the sorts of things which leave themselves permanently exposed to the careful and ongoing scrutiny of a community dedicated to the eradication of all such errors. Whitehead himself spoke of the "formidable permanence of the printed word."[1] But it is exactly this permanence which renders the self-corrective growth of science possible.

Our plan of attack, therefore, is this: In chapter one we will look more closely at the nature of the measurement problem of cosmology, and see how this logical problem need not manifest itself as either a formal or mathematical failure of the theory of relativity. We will begin exploring Whitehead's articulation of this problem in the three works that will be referred to throughout this book as "the Triad," and situate these works within his later, metaphysical theory. An appendix to chapter one will offer an informal survey of some of the mathematical ideas which will appear in our discussion, and also offer some motivation for the informal nature of our approach to what are, after all, formal ideas.

In chapter two we will examine general relativity in some conceptual detail, albeit with only a modest amount of formal mathematics. We will look at the historical and experimental issues that led to the special theory of relativity, including the limitations of this theory and its relative lack of what in mathematics is known as "symmetry," which in turn motivated the development of a more general theory of space, time, and gravity. We will also review some of the historical responses to general relativity, in particular those of Whitehead's, and touch on how his critiques were in turn responded to. We will then briefly examine the current scene in cosmology, and the "cottage industry" of alternative theory production which has emerged since the 1960's.

In chapter three we will take a technical turn, examining more closely how such things as tensors function, and the kinds of structures within which and upon which they operate in the context of general relativity. We will see a little of how these formal structures allow general

1. See (PNK vii) admittedly, in a different context.

relativity to function as a metrical theory – i.e., one that produces numbers which can be compared with observed measurements. But as we will also see, a metrical theory is not a theory of measurement, and with the above details in hand, we will be able to see just how much has been presupposed by contemporary physics in order to interpret these numerical results. This examination will help us justify the claim that metrical theories of gravity do not, by themselves, answer the question, "What is a measurement?" A metrical theory does not by itself suffice to answer what are really *logical* questions about measurement. We will discuss what some of these logical requirements are, especially the vital relation of "congruence."

In chapter four this examination will be continued with a discussion of Whitehead's theory of nature. In the process of describing this it will be shown how Whitehead's theory of measurement is a natural development from his philosophy of nature. However, it will also be argued that the reverse course is possible: namely, one can go from a theory of measurement to a theory of nature. It will be shown how the implicit realist assumptions of Einsteinian theories of gravity regarding the real number line in turn inform those theories of time and space. Whitehead's critiques of these assumptions will then be brought into play to show why they are not tenable. There will be an examination of Whitehead's early theory of extension, showing how this relates to his arguments about congruence relations and measurement. Finally, there will be a brief discussion of some of the later developments of the theory of extension, which lead through *Process and Reality* and into ongoing areas of contemporary formal research.

Chapter five looks at some of the particulars of Whitehead's proposed alternative theory to Einsteinian general relativity. Some mention will be made regarding the problems one encounters when translating Whitehead's formal presentations, due to the outmoded nature of the symbolism used when Whitehead was writing, as well as typographical errors in his text. With those difficulties and caveats in mind, we will examine both his *precis* of the tensor calculus and the specifics of his physical theory. The former will be to gain some sense of Whitehead's own understanding of tensor analysis, and what he thought was important enough to mention to an audience that could not be assumed to be familiar with these ideas. In his physical theory, we will first look at those aspects which correspond to Einstein/Minkowski formulation of special relativity, from which Whitehead's theory is *mathematically* indistinguishable. We will then be able to examine some contemporary responses to Whitehead's theory, and determine if these authors have successfully grasped

Whitehead's underlying philosophical principles, or if they have only responded to the basic physical theory, which Whitehead always saw as an *exempli gratia* of the deeper philosophical matter he was advancing.

The final chapter will take us into some fairly cutting edge issues. A few of the outstanding problems within contemporary cosmology will be mentioned. The existence of such problems from within the standard models of nature and physics indicates that the edifice of general relativity, upon which this cosmology is built, is not without its blemishes. These empirical problems lend additional weight to the epistemological issues presented by Whitehead's critiques of Einsteinian relativity. Three possible avenues of response to these difficulties will suggest themselves. One of these possibilities is that, while Whitehead's criticisms of the measurement issues in general relativity are substantially correct (and ultimately applicable to *any* physical theory), it turns out that within physical cosmology there are formal principles that are sufficient to meet Whitehead's criticisms. As this last avenue seems to be entirely unknown within the literature, we will explore it with some care.

With that said, let us now turn to the details of our project.

CHAPTER ONE
The Measurement Problem of Cosmology

> *In practice, exactness vanishes: the sole problem is, "Does it Work?" But the aim of practice can only be defined by the use of theory; so the question "Does it Work?" is a reference to theory.*
>
> Alfred North Whitehead, "Mathematics and the Good."

I. The Logical Problem with Cosmological Measurements:

Whenever philosophers presume to wander into the territory of a well established scientific discipline, it is reasonable to look upon their claims and arguments with some measure of concern. For it is far too often the case that the fruits of such philosophical dalliances fail to accurately represent the methods, concerns, and conclusions of the science in question. When, moreover, said philosophers take it upon themselves to proffer criticisms of a fundamental nature directed at a scientific field that is noted for its spectacular successes, one can scarcely be blamed for reacting with something akin to outright alarm.

Yet despite these concerns – to a degree, even because of them – such risky business is the defining character of our enterprise here. I propose to engage that centerpiece of contemporary physical cosmology, general relativity, and to do so from the position of a philosophical critic. Now, the opening pages of any extended argument can never be more than a promissory note to the reader, an "IOU" of the argument to come as debt is accumulated in the bald-faced presentation of a problem which is yet to be even recognized as such. This opening chapter is not different in this regard, except perhaps in the extent to which such debt will be summarily, but necessarily accumulated. For the argument to be presented here is quite sweeping in its potential implications for contemporary physics and cosmology. This in turn means that the statement of the issues and the methods which will be used to engage the problems as they are presented here will require no small measure of patience on the part of the reader.

The argument to be made here is that there is a fundamental problem in the standard presentations, conceptualizations, and understandings of all orthodox, or, what I will treat as meaning the same thing, *Einsteinian* formulations of general relativity.[2] This problem makes itself fully

---

[2] For purposes of brevity, I will often refer to this orthodox, Einsteinian formulation as "GR" throughout this work.

manifest in the theory of measurement. The issue to be raised here, stated as blatantly as possible, is that these orthodox presentations of GR render problematic if not outright incoherent the very idea of measurement within cosmology. This "measurement problem of cosmology" is ultimately rooted in the attendant conceptualizations of nature which dominate physical science at large, and finds itself expressed in the interpretations of the mathematical representations of these ideas. However, these conceptualizations and interpretations in their turn are rarely, if ever, made fully explicit in discussions of GR. As a result, this failure to be fully explicit has served to mask crucial issues affecting the logical coherence of the entire theory, thus allowing the problem to fester for decades.[3]

This problem stems from the interlacing issues of the theory of nature, of space and time, and philosophical assumptions that guide the application of our mathematical theories to the above. None of these difficulties directly interfere with the derivation of numerical values from the mathematical structures of GR, nor with the comparison of those derived numbers with the theory-laden observations with which they are associated. But when we pursue the assumed theory of measurement that is intrinsic to GR and its allied theory of nature we discover that, at the very least, there is a substantial logical obscurity buried within these assumptions. Moreover, referring to this problem as a logical obscurity is a kindness; Whitehead claimed it was an outright contradiction.

By this time, the reader should be sufficiently braced for our non-standard use of the term "logical" in these pages to be unsurprised by the discovery that in speaking of a logical obscurity in GR, nothing is being committed toward the formal rigor or transparency (or possible lack thereof) regarding the mathematics of that theory. It is certainly the case that GR presents various mathematical difficulties. Differential geometry and the tensor analysis – which are the heart and soul of GR's formal mechanics – are subjects that are rife with subtleties. Algebraically speaking, tensors admit of a relatively straight-forward understanding, at least at the most elementary level. But even this understanding can be masked by the imposing combinations of upper and lower indices that are exploited by these instruments. Meanwhile, it is well known that the differential equations that are combined within the black-boxes of various

---

3. There are, of course, numerous discussions in the literature on GR, ranging from the straight-forwardly popular to the most technical treatises imaginable, which deal with various levels and layers of the implications of GR's claims regarding the deep structure of physical nature. But these treatments largely take for granted the very issues we most particularly need to question.

tensors of GR often lead to non-linear equations. This latter can, indeed, create a kind of mathematical obscurity, due to the difficulties inherent to finding numerical solutions to such equations.

With regard to some of the above mentioned mathematical issues of GR, the reader is encouraged to refer to the appendix to this chapter. There, a gentle introduction to a few of the formal details are presented in such a way that, while not enough to produce anything like mathematical facility, will hopefully suffice to create a level of familiarity such that one can look at a few of the relevant formulae and see the essential outlines of what these equations express.[4] For the mathematics of GR *is* difficult and subtle. But we cannot permit these difficulties and subtleties to interfere with our close examination of that theory. And even more importantly, we must not under any circumstances permit these difficulties to distract our attention from the problems that we are dealing with here.

General relativity is (among other things) about spatial measurement, and it is on this account that it is logically obscure. Regardless of its mathematical issues, GR is logically obscure because the *possibility* of meaningful measurement is not explained, but largely taken for granted. But the possibility of measurement in GR is something which is dependent upon the logical relations at the center of spatial measurement in general. Yet, as we will see, these relations in turn require circumstances that GR denies.

What kinds of relations are presupposed by spatial measurements? Well, as we have already noted, any measurement requires the comparison of like to like. This means that, for spatial measurement, the comparison of one spatial segment to another is an essential aspect of this kind of measurement. Taking the simplest of circumstances, this means that a spatial segment must be held to be the "standard" to which others will be compared, and this standard must ultimately be manifested in the form of an extended physical object such as a yardstick or a tape measure.[5] It is this physical standard that establishes a baseline of spatial extension that is the necessary basis in any spatial measurement. But as vital as this physical

4. Conversely, readers who are already comfortable with tensors and differential geometry might find the appendix helpful in so far as it situates our informal approach to the formalisms of GR and contemporary cosmology.

5. I will avoid any discussion of the current efforts to build standards that are based upon complex interactions of forces evincing the fundamental constants of nature, as opposed to the traditional, physical object. Eventually, these standards, and the fundamental constants that they employ, must fall back on a spatially extended standard as their baseline of measurement.

standard is, by itself it is not enough.

This physical standard is the source of the similarity that makes comparison possible; this is the core of the "like to like" relation mentioned above. But it is also necessary that it be possible to bring the likeness of the one into comparison with its like in the other. It must be possible to – in some respect or another – bring our yardstick up to that which we intend to measure. If our unit of measure is locked in a vault in the National Institute of Standards in Washington, DC, while the bit of extended space we need to measure is in the form of a 2X4 piece of wood in a lumber yard in Los Angeles, CA, then we must either carry that standard from the one location to the other to make our measurements, or we must have a more readily transported surrogate that we can use for the job. For purposes of measuring yards of wood on this planet, such surrogates are readily found in the forms of the vast array of acceptable measuring devices that permit us to project our standard of measurement (in the NIST vaults) to our various points of interest. This is the second, absolutely essential, relational factor in the logic of measurement. While we must have a standard of spatial comparison, we must also have standard(s) of spatial projection that allow that unit of comparison to be uniformly brought into comparison with – i.e., projected onto – those things we mean to measure.

It is this latter which GR appears to deny us. Because the uniform geometrical relations that make such projective comparisons possible are rendered dubious if not downright impossible by the fundamental assumptions of general relativity. These assumptions most importantly include the idea that the necessary and uniform relations of geometry are collapsed into the contingent physical relations of matter and gravity. This move by Einstein, of identifying physical space with *logical* geometry, has long been viewed by the physics community as one of Einstein's most brilliant postulates. But by collapsing the distinction between geometry as a purely logical discipline and physical space as this is studied in empirical cosmology, Einstein arguably compromised the logical and relational structures which make a coherent theory of measurement possible.

There are two central problems that Einsteinian relativity introduces into cosmological measurements, the first one is serious while the second one is catastrophic. Both problems have to do with the structure of space and time, although I will primarily confine the immediately following remarks to the spatial aspects of the problem.

The first issue is this: in order to eliminate the distinction between physical space and geometry, it is necessary to assume that empirical space is equipped with the same kinds of entities which one postulates within mathematical geometry, so that sense can be made of the proposed relational equivalence of physical reality and geometry. Within GR this assumption takes on the particularly robust form of the presupposition that one can directly identify the empirical structure of space and time with the formal structure of the Real Number line.[6] In particular, it is assumed that space is composed of extensionless points, breadthless lines, depthless planes, and so on. The mathematical term for this kind of relationship is an "isomorphism," and we will be encountering and questioning this assumed "Real Number Isomorphism" (RNI) in much greater detail in later chapters. How one is supposed to justify such a sweeping set of assumptions is quite problematic, for it is far from clear what possible experiential or phenomenological content could be appealed to as grounding such a move.

But this is hardly the only problem faced by the theory of measurement under the assumptions of GR. There is also the matter of how one is to bring into comparison one's chosen standards of measurement and the objective spatial extensions that are to be measured. By identifying the contingent factors of physical (notably, gravitational) nature with the very geometry of space, Einstein eliminated the uniform and necessary relations that make it possible to project our standards of measurement beyond our immediately accessible environs. For according to Einstein, the very geometry of space is contingent and variable, its nature at any given point being dependent upon the influences of matter and energy throughout the universe upon that point.

And here we have the crux of the measurement problem of cosmology. If the very structure of space is a contingent matter of physical influences, then we must first know the nature and distribution of those physical factors before we can know the geometry of any spatial region. But in order to know this distribution of physical factors, we must be able to make accurate and reliable spatial measurements to properly place and relate those factors. But in order to make accurate and reliable spatial measurement, we must have a robust understanding of the geometry of the space that, and through which, we are measuring. Yet such a robust

---

6. Or, in more technical terminology, that space is locally topologically homologous to $\mathbf{R}^n$, where "**R**," or equivalently, "$\mathbf{R}^1$" is the one dimensional real number line, and "$\mathbf{R}^n$" is the n-dimensional continuous space which is, in essence, n-copies of **R**. In this particular instance, the n-copies are not particularly critical. It is the assumed isomorphism between space and the structure of the real numbers, in whatever number of dimensions, which is the issue.

understanding of the geometry of space is precisely what we do not have, and what GR refuses to grant us.

Let us reiterate, from a slightly different angle, how and why this is a problem. When we carry our tape measure from our house to the lumber yard, we are confident that the tape measure continues to mean the same thing at the lumber yard that it meant at the house. Measuring out X number of square feet of plywood, for instance, possesses all the characteristics at the store that we discovered we needed when we first determined how much wood was needed at the house. How is it possible that the tape measure should achieve such a continuity of meaning from the one locale to the other? Well, the tape measure is itself an extended segment of space, and when we carry it from one place to another so as to exploit its characteristics as a measuring instrument, we are asserting that the space itself – at the two locales and as represented within the tape measure – is appropriately comparable. But even if the respective spaces changed in some manner, this would still not be excessively problematic, provided we could actually go to all the respective points in space and determine what our tape measure now meant at each new locale. This would enable us to continue to use our measuring tape, because the rules of its application (although somewhat more complex than if the spaces are all of a kind) would nevertheless be knowable. But on cosmological scales, even on scales only slightly beyond the boundary of our own solar system, we simply do not have this option. We must come up with some reason to believe that our earth based measurements can be legitimately projected to these distant spaces in order to have even the hope of a meaningful cosmology. But if the very geometry of space is something we cannot know until after we can confidently engage in measurement, then cosmology as a science teeters on the brink of nonsense. For while we must first measure before we can know; GR requires that we know before we can measure.

This is the philosophical quandary in which we find ourselves. GR at once saddles physics with a general theory of nature that is dubious if not unintelligible because of its casual acceptance of the Real Number Isomorphism (the RNI hypothesis), and a specific theory regarding the nature of space that renders the very possibility of measurement questionable, because the essential requirements for the possibility of spatial measurement are explicitly denied.

And yet, GR is successfully employed in formulating, making, and evaluating cosmological measurements all the time. Indeed, general relativity and quantum mechanics are often held up as the premier examples of the most successful physical theories ever formulated. How are we to reconcile such practical successes with the supposed philosophical issues raised above?

It is, perhaps, poor salesmanship to speak of such success while simultaneously mentioning general relativity and quantum mechanics, given the well known fact that these two theories are mathematically irreconcilable. However, this is a subject that we shall only briefly touch on in the final chapter of this book. For now it is sufficient to mention two facts. The first of these is one we shall return to quite a bit in the discussion that follows: namely, that there are a large number of alternatives to GR that are equally as successful as physical theories. Moreover, as we shall see, some of these alternatives altogether avoid the measurement problems of GR. Yet, outside of the specialized physics literature, the very existence of such alternatives is all but unknown. One of the problems facing such alternatives is the apparent lack of a philosophical underpinning that could justify taking them seriously. Uncovering the possibility of such an underpinning is a major theme of this book, and an important reason for our interest in Whitehead's thought.

But on a much less abstract level, it must be acknowledged that we are constantly engaged – and *successfully* engaged – in activities that we do not properly understand. Practical success at an activity cannot ever be equated to logical transparency. How many of us genuinely understand the automobiles we drive? How deep must that understanding go before it can be allowed that we posses a "genuine" understanding? Even such trivial matters as crossing the street or catching a ball require the simultaneous solution of systems of differential equations that few of us could even begin to understand. And yet we cross the street and (perhaps less frequently) catch balls under the most casual of circumstances. If we cast even a moment's thought at these homely examples, it is readily apparent that the very ease with which we engage in such activities serves to *obscure* the complexities of the various underlying processes involved in their completion. When we add into our account the fact that there are other theories than GR that present us with numerically precise cosmological predictions, predictions that relate to observable phenomena with accuracy that is comparable to that of GR, then the mere fact of GR's practical success ceases to stand out as such a significant factor in its favor.

We need to ask deeper questions of GR than merely, "does it work?" Such a limited approach might suffice over the short term for physical theory itself, but it is not enough for philosophy. And even in physics, such a sanguine approach can, over the long run, produce more problems than it resolves. The very concept of "working" is itself deeply theory laden, so that claiming that GR "works" is only correct within a limited context that takes little or no account of the *logical* intelligibility of its theory of measurement. The philosophical questions to be engaged here are "meta-questions," if you will, that require us to step back from a full immersion in the methods and practices of physics, so that we can take a stance of critical scrutiny toward those methods. We are not interested so much in learning how to use the tools of physics, as we are in the questions of why and how those tools can (or cannot) serve their intended functions, and what those functions themselves might ultimately mean. Thus, while physicists can, at least for a while, allow the logical issues that underlie their practices to remain implicit, especially when those practices appear to be working so well, it is precisely such fallow ground which is the paramount arena of philosophical inquiry. Thus, the measurement problem of cosmology is a philosophical problem precisely because it stems from the most basic assumptions within physics, assumptions which are all the more difficult to track due to the very successes of the practices within which they hide.

It is at this point that Alfred North Whitehead's work on the philosophy of science and the philosophy of nature offers itself as both an explicit critique of the problems inherent in GR and its assumed theory of nature, and an outline of a way out of these problems. Contemporary with the early development of relativity theory, Whitehead published a series of three books between 1919 and 1922 in which he not only laid out a detailed criticism of Einsteinian general relativity (of which the above is but the barest of outlines), but a systematic approach to nature, physical science, and even a theory of gravity, that could meet the above criticisms head on. These three books were his *Enquiry into the Principles of Natural Knowledge* (1919), *The Concept of Nature* (1919), and *The Principle of Relativity with Applications to Physical Science* (1922).[7] These works will stand at the center of our discussions throughout this book, and will often times be referred to here simply as "the Triad."

The questions Whitehead engaged in these books are precisely the questions we must engage here: What sorts of relational structures – structures of knowledge and or nature – must be presupposed for cosmological measurements to be successful? What sorts of things must be

7. In keeping with standard practice in the Whitehead literature, these works will be individually referred to by the abbreviations "PNK," "CN," and "R," respectively. Other standard abbreviations of Whitehead's major works are set out in the references.

the case in order for these to be meaningful? As he struggled with these questions, Whitehead came to explicitly formulate a theory of gravity that avoided these difficulties. For, as we shall see, it is precisely in the interplay between the nature(s) of space and gravity that the measurement problem of cosmology emerges. Moreover, as we shall see (and as has already been hinted at above), the particular theory that Whitehead presents as an *exempli gratia* is itself representative of an entire class of such theories, many of which can equally well solve the problem Whitehead set out to deal with.

It is worth noting here that Whitehead was well qualified to appreciate GR, both in its strengths and its weaknesses. He was, of course, a leading mathematician of his time. But while he is most commonly remembered today for his collaborative work with Bertrand Russell in that corner stone of contemporary formal logic, the *Principia Mathematica*, Whitehead's career as a mathematician was substantially concerned with problems of geometry and space, and the development of techniques for tackling these problems through logical and algebraic tools. Thus, while his earliest major publication was entitled *A Treatise on Universal Algebra* (Whitehead 1898), a perusal of the major divisions of this volume shows that five of seven "books" are devoted to topics of a specifically geometric character, while the other two are focused on advancing the formal symbolism that will be used in the other five.. His 1906 tract *The Axioms of Projective Geometry* (Whitehead 1906) was cited as recently as 1999 as a standard reference on the subject.[8] And even where his concern was not immediately geometrical in nature, as in his work with Russell on the *Principia*, the concern with matters of space was not far below the surface. It is known, for instance, that there was an intended fourth volume to the *Principia* on the axiomatic development of geometry, which was to have been Whitehead's primary responsibility (See Schilpp 1951, pp 38 ff).

It is also appropriate to bear in mind the scientific context in which Whitehead was working in the years leading up to the publication of his own Triad. Lorentz had been speculating in print on the failure of the Michelson/Morley experiments to detect "ether drift" in the late 19th and early 20th centuries, and by 1902 had already proposed the outlines of those formulae which have come to be known as the Lorentz Transformations. Einstein had sketched the basis of the special theory of relativity by 1905, and Minkowski had set that theory into its contemporary framework of a four-dimensional manifold by 1908. Finally, Einstein had published "The

8. Specifically, in the *CRC Concise Encyclopedia of Mathematics*, "Projective Geometry," Chapman & Hall/CRC, New York, 1999

Foundation of the General Theory of Relativity" in the *Annalen der Physik* in 1916, which set the stage for Eddington's famous 1918 expedition to observe the solar eclipse in the Indian Ocean.[9] Whitehead was well aware of all of these developments, and acknowledges the work of the above scientists and others by name (PNK vi).

Whitehead's criticisms go far deeper than any particular formulae or choice of symbolism of Einstein's, or even any specific physical theory of space and time. Whitehead is challenging the underlying constellation of concepts of the "nature of nature" itself, and the otherwise unexamined intuitions which govern the development of physical theory. Thus, for example, Whitehead does not seek to challenge the empirical predictions of the special theory of relativity at all, stating of his own work that, "[t]he metrical formulae finally arrived at are those of the earlier theory (i.e., Einstein's special theory of relativity)." However, he immediately goes on to add that, "the meanings ascribed to the algebraic symbols are entirely different" (R v). A little later he goes on to express his suspiciousness of the then prevalent (and now all but universal) interpretation of tensors as geometrical entities, and altogether rejects the notion of a "fundamental tensor" so central to orthodox GR, viewing such interpretations as "guises" and "metaphors" which entirely mask the nature of the application Whitehead intended for his own formulations[10] (R vi). So even when the mathematical formulae are altogether equivalent from a practical point of view – practical in the sense that they both render the same numerical predictions – from the larger, *logical* point of view of what the symbols ultimately mean, the differences are sweeping.

Whitehead's development of his criticisms of orthodox GR was no mere irascible or reactionary response to Einstein's radical theories of space and time. Rather, it was a systematic program attempting to come to grips with the fundamental reorganization of the most basic concepts of nature necessitated by the theory of relativity itself. Whitehead's Triad was his attempt to build a philosophical perspective of nature that was at once true to the deliverances of experience, but which fundamentally diverged from the orthodox theory of nature found in GR and throughout physical science in general. Whitehead saw this effort as being irreducibly "amphibian"[11] in nature, possessed at once of an ineliminable philosophical aspect, and not simply a matter of physical science alone. Yet a direct

---

9. See (Einstein, *et al* 1952) for translations of the above mentioned papers.

10. For a brief introduction to the idea of a tensor, I direct the reader's attention to the appendix.

11. This is my term, not Whitehead's.

engagement with the details of physics was also a necessary element of his program. So, on the one hand, Whitehead stated that, "To expect to reorganise our ideas of Time, Space, and Measurement without some discussion which must be ranked as philosophical is to neglect the teaching of history and the inherent probabilities of the subject." But, on the other hand, as he immediately goes on to point out, "no reorganisation of these ideas can command confidence unless it supplies science with added power in the analysis of phenomena." (R 4.) Thus, even though Whitehead spends two-thirds of his time in R presenting an alternative mathematical theory of space and gravity as complete as Einstein's 1916 paper – including, as Einstein did in his paper, an extended discussion on the nature and use of tensors for a scientific audience which, at the time, could not be assumed to be familiar with these tools – for Whitehead this was simply the practical follow through on his philosophical project.

The key idea we will encounter as we trace out Whitehead's ideas is what we shall often refer to as "Whitehead's uniformity criterion." As we wills see, the problems with GR arise because we have no uniform system of relations with which we can project our standard(s) of measurement from the local regions of space where we have something like direct access, to distant areas of the cosmos. We can transport our tape-measures to the local lumber-yards, and reach out to the edge of our solar system with our inter-planetary probes. But inter-stellar – to say nothing of inter-galactic – space lies well outside our immediate grasp. We can study the various radiations that come to us from these distant sources, but we can only surmise what these radiations might mean. However, those surmises collapse into mere wishful thinking if we cannot infer meaningful spatial measurements about both the spatial origins and the intervening distances through which those radiations have traveled. But our estimates of those distances are themselves nothing but wishes and guesses unless we can suppose it reasonable and possible to infer the geometric characteristics of those spaces.

But how are we to do this? How are we to infer the character(s) of spaces we have no direct access to? In particular, if the geometric character of space is altered by every last particle of matter and energy that interacts with that space, as GR tells us must be the case, how can we possibly gain any toe-hold on spatial characteristics on a cosmological scale? Whitehead's solution was that GR is wrong – space, geometrical space, is not subject to the contingencies of physical influences. Rather, space is uniform (although, not necessarily Euclidean), and this uniformity is something to which we have phenomenological access. Because of this uniformity, we have at hand the relational structures necessary to projectively compare spatial extensions that *are* directly accessible to us, out to those that are only indirectly available. It is thus the uniformity of space – the uniformity that GR would deny to us – that makes cosmological measurements possible and meaningful.

II. Process Thought and Whitehead's Philosophy of Nature:

Having said this much, it is important to add right away that our task here is not an exercise in Whitehead scholarship, nor is textual exegesis our principal concern. Such activities are always an invaluable asset to the achievement of any real insight into human culture, knowledge, or (as in our case here) science. And, indeed, we will inevitably engage in such activities, since we are dealing with a well developed arena of ideas: texts and scholarly works comprise the bulk of the materials with which we will be operating. But at no point does the scholarly analysis of Whitehead's Triad, nor the exegetical investigation of his and/or other thinkers' texts, constitute an end in itself for our project. Whitehead's Triad is not the goal of our inquiry, but rather an instrument we will be using to drive our inquiry forward. And like any artisan worthy of the name, it is vitally important that we develop a thorough knowledge and appreciation of our tools. But this is only so that we might use these tools more effectively; they remain just tools.

Still, something must be said here to situate Whitehead's philosophy of nature within the larger project of his metaphysics and process thought in general. While this move will not directly advance the particular argument that will follow, it will nevertheless serve as a way to orient this book and its topic for Whitehead scholars, while persons from other philosophical or scientific backgrounds might gain at least an outline sense of the depth and sweep of Whitehead's thought. Indirectly, our argument might gain a degree of additional plausibility if it can be shown (again, in outline) how it forms an integral part of a systematic whole.

With this in mind, there are three questions that we ought to address in order to situate Whitehead's philosophy of nature. These are, (1) Is Whitehead's Triad (or, specifically, the ideas contained therein) relevant to process thought? (2) Is the Triad necessary for process thought? And (3) Is process thought necessary for the Triad? The answer we will find to each of these questions, albeit with an occasional qualification, is 'yes.' We will deal with each question in turn, but we will deal with them in quite general terms, and will make no attempt to develop even a basic notion of Whitehead's process philosophy. A fuller explication of Whitehead's theory of nature and relativity is a matter for the rest of this book, while any serious analysis of his process metaphysics is entirely beyond the scope of our project.

To the first question then: Is Whitehead's Triad relevant to process thought? After all, the above mentioned problems to the contrary notwithstanding, we have a functioning theory of gravity in the form of general relativity. Does Whitehead's alternative theory offer anything more than an exercise in mathematics? Even more to the point, is this alternative anything other than an exercise? As we shall see in the following chapters, Clifford Will has presented evidence that claims to prove that Whitehead's theory of relativity is empirically false. Given the above, how can these early works of Whitehead's have any bearing on process thought?

As to the first objection raised in the preceding, the logical difficulties with GR that Whitehead has identified cannot simply be set aside. If it was the case that there were no viable alternatives to GR that could meet Whitehead's objections, then the practical necessity of moving cosmology forward with *some* theory of gravity would likely be a sufficient argument to continue using GR. But as we shall discover in chapter five, Will's supposed disconfirmation of Whitehead's theory of gravity is based upon a number of simplifying assumptions regarding the distribution of matter in the galaxy, assumptions that are at least somewhat unrealistic.[12] More importantly, we shall see[13] that there are viable theories of gravity that are every bit as empirically robust as Einsteinian GR. In particular, there are what are known as "bimetric" theories of gravity, so-called because they separate the purely geometric relations from the physical ones. This is what Whitehead argued was a necessary precondition for a coherent theory of measurement. And we shall see that, naming conventions of physics to the contrary, Whitehead's own theory of gravity qualifies as bimetric. It is this quality of being bimetric that makes a theory of gravity Whiteheadian. Thus, the particular viability of Whitehead's theory is a matter of comparatively little consequence. There are other bimetric theories that are clearly viable, and as such Whitehead's general approach to a theory of gravity remains a live option.

It might also be asked, however, if Whitehead himself did not move beyond his earlier philosophy of nature when he wrote *Process and Reality*? For example, Francis Seaman argued for many years that Whitehead explicitly repudiated his earlier work on relativity and the

---

12. Whether or not Whitehead's proposed theory regains viability when realistic assumptions are made is an open question.

13. Primarily in chapters two and five.

philosophy of nature.[14] Again the question presents itself, how can this earlier work of Whitehead's be relevant to process thought?

But this line of questioning seems to be intrinsically weak. We can assert this claim with some confidence because of two facts regarding Whitehead's later work. The first fact is that when Whitehead became convinced that some earlier part of his work was incorrect, he told his readers as much.[15] The second fact is that, when the subjects of physical science and nature come up within the pages of *Process and Reality*, Whitehead's response is the wholesale referral of his readers back to the books of the Triad.[16] Many of Whitehead's self-citations of the Triad are to an entire book in that group, as opposed to this chapter or that section. It is very nearly incomprehensible why Whitehead would make such sweeping references if the ideas of the Triad were ones he believed he had moved beyond. One might still argue that Whitehead was mistaken on this point. But once again, there is scarcely any evidence to suggest that Whitehead was given to such systematic error over the course of many years. Granted there are interpretive difficulties with Whitehead's thought. Yet trying to force this idea to the fore – that Whitehead repudiated his earlier philosophy of nature – ultimately smacks of straining at a gnat while swallowing a camel.

However, it is certainly the case that Whitehead's thought moved forward, and this subject provides us with the segue to our second question, is the Triad *necessary* for process thought? This question is more difficult than the other two, and the answer that is given here will in turn be much more controversial. Since we cannot pursue the full technicalities involved with this issue, our answer will be the more controversial for its brevity.

In any event, our answer to the second question is also "yes." The reason the Triad is necessary for process thought can be stated thus: The Triad is about the nature of time, while *Process and Reality* is about the relational structures of becoming. The choice of words here is not

---

14. See, for example, (Seaman 1955) and (Seaman 1975). We will not take the time here to explicitly critique Seaman's argument.

15. We will explore this fact in more detail in section iv of chapter 4, particularly as this relates to aspects of Whitehead's theory that have taken on a foundational role in certain contemporary arenas of formal research into spatial relations.

16. Thus, for example, see PR, the footnotes on pages 125, 243, 333; also, see SMW the footnotes and text on pages 121, 124, and 158, and AI, footnote on page 254. For examples of Whitehead correcting himself, see PR the footnote on page 128, and pages 287 ff.

accidental. In characterizing the Triad as being about the *nature* of time, it is emphatically the case that time in these works is considered from the standpoint of physical nature and natural science. The focus of the Triad is specifically *not* that of metaphysics; Whitehead calls it "pan-physics" or philosophy of science, more or less interchangeably (R 4 – 5).

On the other hand, PR *is* doing metaphysics and as such is not concerned with time as this is found in nature or described in science, but rather with the relational structures that are the possibility for time so construed. Indeed, the third and fourth parts of PR ("The Theory of Prehensions" and "The Theory of Extension") are devoted to the development of an interlaced theory of internal and external relations, respectively. Thus, for example, part III of PR (oftentimes referred to as the "genetic account") gives us a thoroughly holistic and non-linear account of the internal relatedness of becoming.[17] Meanwhile, time and space – which is to say, physical time and physical space – are products and particular specifications of extension. Physical time, in particular, "expresses the reflection of genetic divisibility into coordinate divisibility." (PR 289.) Physical time is something which presupposes, and in that sense comes after, the relational structures of prehension and extension.

However, this deeper analysis of the relational structures involved does not obviate the need for an account of physical time and space. Quite the contrary, the relational structures of PR presuppose those of the Triad as that which they are (logically) building towards. Nothing in PR gives us the *natural* connection between time and space. Indeed, it is precisely when this subject comes up that Whitehead most casually redirects his reader's attention to his previous work (for instance, PR 333). As we will see below, this connectedness between time and space is an intricate one, and central to it is what will be called Whitehead's "multi-threaded" theory of (physical) time.

Conversely, the above also answers our third question, is process thought necessary for the Triad? Clearly the answer is "yes," for the same reasons of mutual support and presupposition. But the case is much clearer with the Triad's presupposition of process thought. Had Whitehead never written PR, the triad would remain an expressly process-oriented collection of books. Throughout the Triad, the ontology is explicitly one of *events*. Objects and places, particular moments of time or points of space, these are all derivative abstractions from the basic ontology of events, and the

---

17. Elizabeth Kraus has done a particular service to Whiteheadian scholarship by carefully diagramming the relational structures of the genetic account. Note, in particular, the multiple "feedback" loops in her flowchart (Kraus 1998, 125).

extensive relations that manifest themselves as nature.

It might reasonably be argued that, while process thought of some kind is essential for the Triad, *Process and Reality* is not. This is not a question of any great import for us. There may indeed be other kinds of process metaphysics that would serve as a logical ground for the Triad. Certainly it is the case that the only such ground that can work, given the exclusively event oriented nature of Triad, will still be some kind of process system. But conversely, any such process system that might serve to ground the Triad will have to be one that builds up a system of internal and external relations such that metaphysical questions of the one and the many can be successfully addressed, while the relations of extension necessary for the theory of space and time are also adumbrated. *Process and Reality* does all of this. Indeed, it is arguable (although we will not attempt that argument here) that *Process and Reality* was written precisely to resolve the metaphysical issues (the one vs. the many; mind and matter; internal and external relations) that the Triad had left unresolved.

Finally, it might be supposed that Whitehead's philosophy of nature, for all of its talk of an event-based ontology, might yet be compatible with a more orthodox view of nature, such as that which is implicit within most of physical science. However, such a supposition can only seem possible so long as one has not actually read Whitehead's texts. While it is true that the first, full critique of orthodox, scientific materialism that one finds in Whitehead's work, that touches upon both its historical and contemporary aspects, appears in *Science and the Modern World*. Yet that critique is also evident in the Triad. It is somewhat more implicit, but only somewhat. Once again, if we read the Triad as part of the systematic whole of Whitehead's thought, we find a coherent structure that, while admittedly in its earlier stages, is nevertheless continuous with that which follows.

We have covered with speed a great deal of material here, and it might be worth our time to review that material, as it frames the task we have set out for ourselves. It is our task here to engage a central problem within cosmology which cuts to the very heart of the possibility of performing meaningful measurements. At the same time, we are not going to be *doing* physical cosmology. We are going to be doing philosophy, and using those philosophical works of Whitehead which I am calling the Triad as some of the primary instruments of our critique. But we are not going to be doing Whiteheadian exegesis either. Rather, we are using Whitehead's ideas to make explicit the logical problems underlying the measurement problem of cosmology. This problem stems from the failure of GR to separate contingent physical factors from the needed uniformities of pure geometry, uniformities which make it (logically) possible to project our measurement relations to otherwise inaccessible regions of space. It is only this latter fact, made evident by Whitehead and which we will call his "criterion of uniformity," that allows a measurement to take on even the possibility of cosmological significance. Finally, while nothing more will be said on the matter in this book, we have at least seen the outline of how the works of the Triad fit in with Whitehead's later, metaphysical works, and with process thought at large.

An appendix follows that sketches the nature of our informal approach to the formal mathematics of GR and other theories of gravity. Anyone who is not familiar with differential geometry is encouraged to read this appendix, to gain some sense of the material that follow. Anyone who is already familiar with this area of mathematics is encouraged to read the appendix to gain some sense of how we will be approaching these topics. From there, we will move on to the body of our argument itself.

III. Logic and Inquiry:

A few more words need to be said now regarding our use of the term "logic" and its relation to the theory of inquiry. As already stated, it must not be supposed that in challenging the *logical* coherence of GR, any criticism is proffered regarding that theory's *formal* or mathematical rigor, or that the soundness of the numerical results deduced from GR's machinery is in any way flawed. The confusion that presses itself upon us stems from the use of the terms "logic" and "logical." The notion of "logic" and "logical" that is being employed here is significantly larger than the unfortunately narrow meaning those terms have taken on within contemporary discourse. People have, for the most part, come to think of "logic" as referring exclusively to the formal manipulation of symbols, to systems of deductive relations, and the rules of validity which might be shown to hold within such systems. Certainly such relational structures are of vital importance in any discussion of, or use relating to, logic. But they are not the only things of importance.

In particular, there is an older notion of logic that is rarely – though still occasionally – invoked, that treats logic as dealing with the general theory of inquiry. Formal rules of deduction continue to be matters of central concern. But the concern in this case is instrumental to the larger project of logic, where logic is viewed as inquiry into inquiry *per se*. Logic in this view begins as erotetic logic, and the study of the formal relations of necessity, validity, and deduction are handmaidens to this more primary function. It is with respect to this project that measurement within GR appears as a problem, and as a specifically *logical* problem to boot.

Still, the question presents itself, why speak of this as a "logical" problem at all? It might well seem that there are other terms in current usage that do all the work we need done without inviting confusion, such as the proposed use of the term "logic" does. For example, at first blush the term "methodology" exactly covers the matters we are attempting to address. So why not use the term "methodology" here, rather than "logic"?

There are several responses to this query. In the first place, the common associations of the term "methodology" are every bit as problematic as those of "logic." The former term is at once too broad in some respects, and too narrow in others. On the one hand, "methodology" is too broad because it can frankly mean *anything* what-so-ever. The term carries no essential historical or formal connection with the idea of inquiry. Any collection of habits and intuitions that "work" can constitute a method. But it is only as we turn to the logic of that method that we inquire into why and how that method works, to the degree that it does. On the other hand, "methodology" is too narrow because it has left behind the vitally important aspect of the formal and necessary character of the relations we are investigating here. As will become apparent in the development of this investigation, there are conventionally logical problems of consistency and coherence in the very heart of GR.

But the argument for the use of the term "logical" does not stop there. It turns out that there are historical and systematic reasons why the term "logic" should be given preference here. With regard to the first of these points, the concept of logic as a theory of inquiry has both historical roots and contemporary defenders. The roots of this approach to logic go back at least as far as Aristotle's explicit codifications of logic in his *Organon*, though we can see the images of it in Plato's dialogical and erotetic methods of philosophy. With respect to Aristotle's work, it is in the *Topics* that Aristotle most fully develops the inquiriential nature of logic, as the hand-maiden to the art of interrogative reasoning. On the other hand, while the full scope of logic as inquiry might be missed if one focused too exclusively on just the *prior* and/or *posterior analytics*, even here the erotetic nature of logic – the approach to logic as an engine of inquiry – is at least arguably present.

The contemporary thinker who has most forcefully made this claim is Jaakko Hintikka. Hintikka's credentials as a formal logician are impeccable. This in turn lends considerable weight to his arguments when he takes it upon himself to criticize the orthodox, contemporary approach to the teaching and the meta-theory[18] of logic as fundamentally flawed. Logic, Hintikka argues, needs to return to its roots as the study in the art of good reasoning, rather than just the abstract and often mechanical manipulation of symbols. Formal methods of deduction and proof, tests of validity, and so on, continue to occupy a place of importance. But that importance is dictated by the service these techniques can render to the larger topic of good reasoning. But good reasoning is always motivated: as such, it is interrogative and erotetic; it is inquiry.[19]

A third advocate of such an approach was the American pragmatist, John Dewey. For Dewey, logic simply was The Theory of Inquiry. An "old school" philosopher who was educated and matured well before the differences between "analytic" and "continental" styles of philosophy emerged and were named, Dewey witnessed with considerable disfavor the way in which the fascination with purely formal matters was eclipsing the more original sense of logic that he advocated as the true matter of logic: the inquiry into inquiry itself. For Dewey, Aristotle's logic is less a matter of formal deduction than it is a schema of *knowledge* – knowledge as this was conceived by the Greeks of that day. But similarly to Hintikka, Dewey saw the failure to understand Aristotle's logic in the fullness of its context and intended usages resulted in reducing this older system to the sterility of a purely formal system.[20] Formal matters are important, of course, and to the extent that such things are, they are also of interest in their own right. But the primary function of such machinery is to clarify and make explicit meanings and relations as these emerge in the course of inquiry.

So what is this "more than merely formal" aspect of logic that these two thinkers have independently advocated, and how does it come into play here? Most of the factors needed for our work have already been mentioned or alluded to in the above. But for the purposes of clarity, let us collect these ideas together:

- Logic as inquiry goes far beyond the contemporary view of logic as the purely formal manipulation of symbols. Such formal and systematic work is of enormous value in clarifying the relations of implication

18. Since it leads to no difficulties here, I will make no effort to differentiate between logic and its meta-theory.

19. Many of Hintikka's papers on this subject have been collected together in the fifth volume of his "Selected Essays." See (Hintikka 1999) for this, but also the more recent (Hintikka 2001).

5. See (Dewey 1938, pg. 85, and chapter 5.) It should be noted here that Hintikka does not appear to be aware of Dewey's much earlier arguments.

between ideas and propositions. But while such puzzles can be of mathematical interest in their own right, their principle job is in serving the larger erotetic project of asking questions, knowing what questions we are really asking, and thereby having a serviceable grasp on what might, or might not, count as an answer.

- Logic as inquiry is always situated. It is always embedded in a context that is not just or even primarily abstract. Rather this situatedness also includes factors that are cultural, biological, physical, and so on. Questions are posed neither blindly nor at random, nor are they posed in the absence of a baseline of the results of prior inquiries. These topics also comprise the subject matter of logic as inquiry.
- Why call this project "logic"? Because until the last 100 years, this is how the term and the discipline was essentially understood. Rather than trying to invent a term to fill an old meaning, we will discipline ourselves to revive the old term in its original meaning.

It is also important to note that the theory of inquiry is not the same as epistemology. The theory of inquiry is just that: a theory of *inquiry*, not knowledge. Ideally, knowledge is what results when the inquiry is successful. But knowledge is just that – a result. In this regard, the difference between logic and epistemology is the difference between process and product. Inquiry is something we do, and the kinds of doings that we can engage in under the aegis of inquiry are substantial. Some of these doings are unwise; others are perhaps more appropriately founded given the nature of their subject matter. The purpose of *logic*, as we are using this term here, is to make these distinctions clearer.

With these ideas in hand, it becomes easier to understand how the theory of measurement might be a logical matter. Measurement is a vital part not only of scientific inquiries, but many everyday ones as well. What must be presupposed in order for a measurement to be successful? What kinds of things must be the case for a number-generating activity to properly qualify as a measurement? What sorts of relational structures must be in place to enable us to project those numbers onto the world in an effective and meaningful way? All forms of measurement would seem to involve some kind of comparison, and in particular a comparison between that which is "in hand" (our measurement device) and something else which is to be measured. But how many ways might there be to "legitimately" make such comparisons? Any such comparison would seem to necessitate bringing together two things which share some kind of fundamental similarity. But is there ever only one sort of fundamental similarity? If not, which kind is the appropriate one to use in the given comparison?

All of the above questions are vital matters for a theory of measurement, and all of them are of an essentially *logical* character. When the subject matter becomes spatial measurement, some additional characters of interest emerge, most particularly those of extension and the comparison of extensive wholes with one another. When we inquire more closely into the applications of the relational structures of extension, we discover a broader field of employment than just that of space. But once again, because this inquiry is directed toward the fundamental factors of measurement *per se*, it falls under the umbrella heading of inquiry into inquiry. It is, again, a matter of logic.

Within this vein, a word should be said about mathematical inquiry itself. As already noted in the first bullet point above, formal and mathematical inquiries are often pursued for their own sakes, as well as for their usefulness in applications to other types of inquiry. As inquiries, such investigations provide ample areas of study for logic – again, most particularly as that term is to be used here. Above and beyond such insights that symbolic logic might provide to the underlying relations in proofs and mathematical models, logic as the theory of inquiry has much more to tell us about the process of investigating mathematical ideas than just the *post hoc* formalization of the results of such investigations. For these mathematical inquiries are themselves influenced – and often enough, governed – by factors which are themselves more than just formal matters of axiomatization or proof. There are what can only be called the aesthetic characteristics of the relations and structures under investigation, with mathematicians notably drawn to those aspects of their subject which are identified as the most beautiful. There is the seldom discussed nature of mathematical intuition itself, which so often determines the nature and direction of inquiries.

There is, moreover, the indisputable fact that mathematics is not "self-interpreting." By this last I do not mean that appropriate model-theoretic structures never present themselves as "interpretations" of some other formal system. Rather, in saying that mathematics is not self-interpreting I mean that there is no method of relating a particular formula – or, more properly, system of formulae – to the objective world that is somehow already built into the mathematical formulae themselves. This relating of the one to the other is something that *we* as inquirers do; it is not forced upon us by the mathematics itself. In the case of model theory, we are relating one abstract structure to the other. But both of these structures are ours. In the case of the objective world, we have not the liberty to legislate what it is “really” like, the way we have with a mathematical model, nor can we suppose that we already know how the world is to be directly related to our formulae.

We will be exploring these aesthetic and interpretive issues in greater detail in the following chapters, as well as the appendix to this chapter. It is sufficient for now to note these issues are here, and recognize how these and other complexities fall within the scope of a theory of inquiry, what we will speak of as the logical issues surrounding the measurement problem.

Appendix:
Who's Afraid of Differential Geometry?

At this point it is necessary to change directions somewhat. We need to spend some time both examining and exemplifying the methodological approach we will adopt toward the formal machinery of physical cosmology. Our project here requires that we have a working understanding of GR. But this can only happen once we have also gained a working familiarity with the basic "language" with which that theory is set out: to whit, differential geometry. But the "working" understanding we need to develop here is ultimately not such an intimidating thing as might be first supposed. As we shall see, our purpose is not to achieve any level of mastery of the relevant mathematics as a deductive technique. Rather, our working understanding requires us only to gain an appreciation for the mathematics as a text with which we are unfamiliar. To gain this kind of appreciation, we will have to examine a little of what that mathematical symbolism is intended to achieve, and then bring this process to life with some actual mathematical examples. Along the way, we will also enrich our methodological survey with a discussion of the processes of abstraction and analogy as these function in formal inquiries.

It was Whitehead's observation that, "it certainly is a nuisance for philosophers to be worried with applied mathematics, and for mathematicians to be saddled with philosophy" (R 4). Were anyone else to say such a thing besides Alfred North Whitehead (who was evidently unencumbered by even the tiniest shred of irony) we might justifiably wonder if we were not being mocked. With the exception of that small group blessed with a love of and predilection for things mathematical, the general rule for most people seems to be, when confronted with any formula or set of symbols beyond what one might employ while reconciling one's checking account, withdraw with as little outward display of panic as one's dignity is capable of managing. Our challenge here is to somehow get past this unfortunate, but all too common dread of formally systematized ideas, and develop a measure of sanguine acceptance in the face of relational structures explicitly scripted in those harmoniously abbreviated systems of symbols we call mathematics. Stated more simply, we must find a way to treat informally of formal ideas.

Our task is rather akin to that of an expert in the Chinese language endeavoring to present an untranslatable pun in that language that turns on a minor difference between two written characters, to an English speaking audience without any other training or experience with the subtleties of Chinese calligraphy. Or, alternatively, one might think of our situation as that faced by a musician explaining to an appreciative but untrained audience the subtleties of musical notation, what it makes possible and what it forbids, and the kinds of openings for interpretation such notation provides for the musically adept practitioner.

A similar task faces us here. For as it was pointed out above, the practical successes of GR in producing theoretical predictions, couched in terms of numerical values that are eventually related to measurable quantities, does not of itself assure us that the logical basis of those practices has been put on a sound footing. We cannot hope the discussion within these pages is sufficient to make us mathematical *practitioners* ourselves. Nevertheless, we might legitimately hope to gain a degree of *appreciation* for mathematical notation, in order to get some kind of sense for what it makes possible and what it forbids, and the kinds of openings for interpretation such notation provides. I submit that such an appreciation is as achievable in the case of mathematics as it is in the previous examples of Chinese or music. What is called for here is not decades of painful study, only the willingness to set aside a habitual dread of the unfamiliar, and the same openness to being exposed to new names and strange forms that one would bring to, as another example, an art opening.

But at the same time, we cannot be flip or blasé about our approach to this material. General Relativity is a mathematical theory which can be descriptively characterized in natural language with a greater or lesser degree of adequacy (depending on the particular gifts of the person doing the characterization), but which cannot be seen in its own light without making a very real, if only minimal, effort to grasp the formal systems of symbols with which that theory is expressed. We may agree with Whitehead that, "There can be no true physical science which looks first to mathematics for the provision of a conceptual model," (R 39) and yet still be obliged to assert (as, indeed, Whitehead also did) that the precision achieved in reasoning about time, space, rates of change, and the practical aspects of measurement; the elaboration of sound, deductive reasoning; a refined theory of ordinal relations; and so on, all render the contributions of mathematics to physical theory absolutely vital (e.g., PNK v). Thus, if we are to grasp how the mathematical machinery of GR ultimately *fails* to satisfy the logical requirements of a theory of measurement, we must be prepared to brave at least the outskirts of this jungle of alien symbolisms.

Now, however complicated a mathematical formula or set of

symbols might appear to the uninitiated, the ultimate aim of all such representations is the *simplification* of the process of expressing an idea, an idea which is itself part of a context and constellation of other ideas. Each symbol is essentially a substitute for a more complicated set of signs, which themselves are often enough just compressed versions of still more complicated signs. Eventually a set of formal symbols can be translated back into something like ordinary language, but the expression thus achieved would itself be so cumbersome and convoluted, so impenetrably opaque in its complexity of articulation and so ghastly in the inelegance of its phrasing, that it would scarcely be possible for the human mind to understand its meanings, much less reason effectively from them. By substituting new symbols for the more complicated expressions, symbols divested of any unnecessary conceptual baggage that would lead to ambiguities, enormous economy of thought is achieved (for instance, see Whitehead 1898, pp. 3 ff).

Thus, for example, the "tensor" idea mentioned above (pg. 16), which is an absolutely vital instrument in the expression of all contemporary theories of space and gravity, is ultimately nothing more than a book-keeping device. It is simply a method of keeping track of other, lengthier equations (which, for all of that, are not especially mysterious in their own right) through the use of various combinations of upper and lower indices. Thus, things like "T^{xy}" or "$g_{\mu\nu}$" or "R^{i}_{jkl}" would be examples of tensors, where the regular letter names the tensor, and the sub- or super-scripted indices perform the book-keeping work. Depending, of course, on the context of their use, such symbols might mean a wide variety of very different things. But within the context of the tensors themselves, there are a great many things that can be done with them, all without fully grasping the underlying meanings of the equations for which the tensors themselves are the compressed expressions.

Herein lies both our curse and our salvation. The curse is that we find ourselves staring at a "black box" whose inner workings race beyond any evident hope of our comprehension. This may indeed be a major source of the "math-phobia" which seems so common. The very triumph of mathematical symbolism – namely, its economy of thought, exactness of expression, and self-contained purity – also serves to shroud these symbols in an almost arcane air of total mystery. It is as though there were no hope of penetrating to the bottom of the depths of their meanings. Nor is such dread of the abyssal plumblessness of mathematical arcana entirely limited to the "uninitiated." Even amongst mathematicians, there is a tendency to react with a disciplined caution, whose professionalism masks no small measure of trepidation, when confronted with mathematical ideas which lie outside of the particular individual's immediate technical education and expertise.

The reason for this is quite simple: There is just too much mathematics for any one person to master. A contemporary (1979) estimate of the sub-divisions of areas of specialization within mathematics came up with over 3,400 sub-categories. It has also been calculated that there are somewhere in the neighborhood of 200,000 theorems published every year, a number vastly greater than any one person could ever study with any degree of comprehension (Davis and Hersh 1981, 17 – 30). Specialization within mathematics has reached such an extreme that an individual's primary area of research may only be comprehensible to a "few dozen or at most a few hundred" others in the same field worldwide (Davis and Hersh 34). Indeed, so vast is the range, complexity, and variety of forms of mathematical inquiry that the very notion of mathematics as a single, unified and uniform discipline has been characterized as a "myth" (Hersh 1997, 37 – 8).

The point of all this is not to despair of ever learning enough mathematics to properly grasp the inner workings of GR. Quite the contrary! We can now rejoice at the realization that we need not learn all of mathematics in order to learn a little of it. Here, in turn, lies the blessing of mathematical symbolism. Since all that anyone ever knows of mathematics is but a little of it, we need only burden ourselves with so much of the subject as will efficiently serve our purposes. And for this purpose, all we need is a summary glance at the operations of the requisite mathematical machinery. This the key to the statement made above, that what we want is an informal grasp of formal ideas.

We are interested in becoming intelligent and informed critics of contemporary cosmology. And, in this context, "critic" must be understood in the most constructive sense possible, as "One skillful in judging of the qualities and merits of literary or artistic works."[21] In our case, the literature whose qualities we are engaged in judging happens to contain a substantial amount of description of physical phenomena in the form of mathematical symbols and equations. But the need for refined judgment which is nevertheless external to the matter being judged is not the less real because our subject falls under the heading of science. Our philosophical aim here, "then appears as a criticism and a corrective," (R 5) to the inadequacies of the philosophy of nature which has dominated the development of GR over the past century. It is worth adding, in this

---

21. Quoted from the *Compact Oxford English Dictionary, Second Edition.*

context, that the kind of "criticism" we are engaged in here is neither passive nor merely reactive to the data of science. The continuation of the above quote from R reads: "and – what is now to the purpose – as an additional source of evidence in times of fundamental reorganisation." The value of criticism is severely limited unless it takes the constructive step of actually offering new insights that can be built upon, as opposed to merely pointing out the insufficiencies of the existing ideas on a subject. This, of course, requires a familiarity with the tools of GR. But such familiarity is not the same as what might be called "facility." It is obviously possible for a person to know how to drive a car without being able to build or even repair one. Indeed, it is even possible for a person who does not know how to drive to nevertheless recognize the skillful operation of a vehicle – there are, for instance, far more appreciative observers of the Navy's Blue Angels than there are jet pilots of any level of skill. Such is our position towards the mathematics of space, time, and gravity.

We do not need to operate the machinery[22] of GR at any significant level of skill. We do not need to dread the appearance in the text of a tensor, any more than we are angst ridden in the presence of a machine whose inner workings we scarcely comprehend. What is important to keep in mind is that whatever subject area one is familiar with, be it mathematics, technology, art, language, history, etc., there is always a "something more" involved which is beyond one's comprehension, and which renders the subject at hand a kind of "black box." We find ourselves in the midst of the unknown all the time. However uncanny it might seem when we encounter it, whether it be a jet aircraft, a cubist painting, a jazz improvisation, a pun in Chinese, or a differential equation, the unknown is not an antagonist confronting us, but an opportunity inviting us onward.

In our case here, we need to be able to look at such things as tensors, differential equations, and other pieces of mathematical exotica. We need to look at them carefully, discriminatingly, and with the courage of our convictions to challenge these symbols to provide us with answers to our questions. We cannot permit the mere mathematical form of an idea to intimidate us into accepting it at face value, simply because it has assumed the venerated form of mathematical symbolism. If a collection of mathematical symbols is to be interpreted as making a claim that some set of relations factually hold in the world, then even if we cannot deftly

---

22. It should be mentioned that the use of machine metaphors here is not an arbitrary choice. Throughout their work on GR, Misner, Thorne and Wheeler constantly refer to tensors as kinds of machines (Misner, Thorne and Wheeler 1973, pg.22 ff, and throughout).

operate the deductive machinery of the formula, we may still expect to be able to get some sense of *why* the formula may be said to make the claims it does. We can look at a symbol, and compare it to one that is similar without being identical, and know something of what the differences mean. We can learn to look at something like "T_{ik}" and compare this to "T_{ki}," and see that they are slightly different symbols.[23] We can know that if they say the same thing, then we are dealing with a "symmetric" tensor, and know that "symmetry" tells us something of vital importance. On the other hand, if we are told that the two say something slightly different, we can see this fact in the expression

$$T_{ik} - T_{ki} \neq 0$$

and know that equations much like this hold the key to the claim in GR that space is curved. This can all be achieved without ever fully grasping the inner workings of either "T_{ik}" or "T_{ki}," such as would enable us to actually deduce numerical values from them.

As a matter of fact, our tensorial friends above are intended as abbreviations for differential equations. These are, in turn, given geometrical interpretations within GR, and can also be studied in their own right as they relate to concepts of space. When this is done, one has that branch of mathematics known as "differential geometry." Now, I Have used this term "differential" on more than one occasion without any sort of explanation. It is time to rectify that problem, and do so in a fashion that will exemplify in a more concrete manner some of how this "informal grasp of formal ideas" will work.

To begin with, let us just look at the name, "differential geometry." Focusing on the names of mathematical entities is not always very helpful (is a "tensor" just under too much stress?), but in our case here we can pull out some serviceable information. Presumably, the term "geometry" is obvious enough to all; it means the formal study of space, where the notion of "space" is itself left quite open, but at least somewhat rooted in ordinary notions which we ultimately owe to the direct deliverances of experience. The qualifying term "differential" is most likely the greatest source of uncertainty. But the root term "difference" is again a solid clue.[24] Differential equations relate the effects of small changes – i.e., differences

23. Specifically, note that the subscripts are reversed between the two.

24. However, it is also important in this context to note that there is something known as a "difference equation," which is related to, but certainly not the same thing as, a differential equation.

– within some system or structure. Thus, “differential geometry” is the study of how the structure of a space can change from one point to the next. So far, so good. But how does that differential go about doing the kind of work it is expected to perform?

In order to give these workings a casual glance, we will need to learn a few more technical symbols. This brings us to one of the main hero's of this part of our discussion, Silvanus P. Thompson, whose classic little work *Calculus Made Easy* stands as an exemplar of effective pedagogical clarity. Thompson opens his delightful, if somewhat iconoclastic tract with the following observation: “Considering how many fools can calculate, it is surprising that it should be thought either a difficult or tedious task for any other fool to learn how to master the same tricks” (Thompson 1987, vi). Born in 1851, educated as an engineer, and a Fellow of the Royal Society, Thompson quite evidently lost all patience with what he viewed as the almost gratuitous obfuscation of the teaching methods of his day. His book, originally published in 1910, was the result. Thompson's introduction to the ideas of the differential and integral calculus is worth quoting at some length here:

> The preliminary terror, which chokes off most fifth-form boys from even attempting to learn how to calculate, can be abolished once for all by simply stating what is the meaning – in common-sense terms – of the two principal symbols that are used in calculating.
>
> These dreadful symbols are:
>
> (1) *d* which merely means “a little bit of”.
>
> Thus, *dx* means a little bit of x; or *du* means a little bit of u.... [Y]ou will find that these little bits ... may be indefinitely small.
>
> (2) ∫ which is merely a long S, and may be called (if you like) “the sum of”.
>
> Thus ∫*dx* means the sum of all the little bits of x; or ∫*dt* means all the little bits of t. Ordinary mathematicians call this symbol “the integral of.” Now any fool can see that if x is considered as made up of a lot of little bits, each of which is called *dx*, if you add them all up together you get the sum of all the *dx*'s (which is the same thing as the whole of x). The word “integral” simply means “the whole”. ...
>
> When you see an expression that begins with this terrifying symbol, you will henceforth know that it is put there merely to give you instructions that you are now to perform the operation (if you can) of totalling up all the little bits that are indicated by the symbols that follow.
>
> That is all (Thompson, 1987, 1 – 2).

Indeed.

Now, an ordinary differential is a relation of the change of some one thing in respect to another individual thing. If the first of these is "x" and the second is "y", then the differential of x with respect to y would be written as dx/dy. It is important to keep in mind that the "d" here is not a number that is multiplied with x or y; rather it, like the integral, is an operator, a machine (to fall back on our earlier metaphor) which takes that x or y, and returns "a little bit" of it – a very slight difference. In this case, it is the slight difference of x that occurs when there's a tiny difference made in y.

Notice also that, so far at least, we have only permitted the change of one thing in respect to one other thing. But we want to be able to apply the idea of a differential to geometry, to space. From any given point in a space, things might be changing in a great variety of different ways, depending on the point one at which one is looking, and the direction to and from which those changes are presenting themselves. What happens when there are many "little bits" to be changed at once? For this we need a refinement on the idea of ordinary differentiation.

Let us once again make things relatively concrete. For purposes of an example, let us say now that our "x" is going to vary in respect to both "u" and "v." Now, we know how to make x vary with respect to just u, if only we could pretend that the v played no significant roll in how x varied. This is simply our dx/du from above, where we are using a "u" instead of a "y." By the same token, we know how to make x vary with respect to only v, since this is just like the case with u (i.e., dx/dv). So, rather than trying to figure out what happens when we vary u and v at the same time, we can treat first the one and then the other as fixed (the rules for handling such fixed or constant values are quite straight-forward), and vary (that is, differentiate) x with respect to just one part of the whole at a time. This *partial* differential is exactly the trick we are looking for.

It should come as no particular surprise to anyone that this partial differential has its own symbol, which in this case is "∂." The rules for calculating values of partial differential equations (known as "p.d.e.'s") can be quite a bit more complicated than their ordinary ("o.d.e.") siblings. Happily, few of these subtleties will require any of our attention. What we need to know for now, is that whenever we are confronted by a regular differential symbol, such as dx/dy, we are dealing with a problem which can basically be solved in a single stroke. When, however, we are faced with something like $\partial x/\partial u$ and/or $\partial x/\partial v$, we are only working with a part of the problem at a time, because the functional relations between x on the one and, and (u,v) on the other, are too complex to permit of a direct solution.

One can also observe that there is a bit of redundancy in the symbolism of $\partial x/\partial u$ and $\partial x/\partial v$. Mathematicians hate writing a lot when they can get away with only writing a little. So, to begin with, if we remove the "x" from both $\partial x/\partial u$ and $\partial x/\partial v$, we no longer have a partial differential that could give us the value of x in relation to u or v, but we do have a couple of "operator symbols" ($\partial/\partial u$ and $\partial/\partial v$) which are of interest in their own right. A symbol such as $\partial/\partial u$ need no longer be applied only to x. It can now be generalized to other possible relata, and its functional characteristics studied in their own right. And still we can abbreviate some of the symbols into something more compact. We frequently find $\partial/\partial u$ written simply as ∂_u and thereby represent the same functional operator with less redundancy – and, of course, a similar move with our other differential operator gives us ∂_v.

We will certainly see more of these differential operators, and similar mathematical devices that have evolved from the basis of standard p.d.e.'s. This little excursus into differential functions and operators scarcely manages to scratch the surface of the extent to which a profoundly simple idea – take "a little bit" of something, and study the differences that "little bit" makes when it varies – has been built, refined, and made progressively more rigorous. But it may be hoped that this very brief tour will serve both to indicate the intuitive and informal level at which we will be handling the mathematics of GR, while at the same time discovering how simple but important ideas present themselves in what might otherwise appear to be the pretentious formalism in which these ideas are clothed.

In our instance here, we have already seen that the seemingly trivial notion of "a little bit" of change manages to break itself apart into two major kinds: the ordinary differential which handles a problem as a single whole and resolves that problem in what amounts to a single stroke, and the partial differential which breaks the problem of "a little bit" of change into many such "little bits," solving them one part at a time. This latter approach naturally lends itself to geometrical problems, where the *multi*-dimensionality of space is a congenial interpretation of, and justification for, the *many* "little bits" that must be varied independently in order to resolve the differential problem at hand. Each dimension offers one family of "little bits" to be varied from any particular point, from which we can work out the structural characteristics of that space as one moves from one point to another, a "little bit" at a time. The partial differential is not quite the final tool we need to complete this analysis, though: we require something called "covariant differentiation," which is built out of the basic

P.D.E. with some "corrective factors" added in. A discussion of those specifics can be postponed until chapter three. For now, the intuitive ideas here are enough to enable us to move forward. It turns out that it is in the order in which these many "little bits" are dealt with (using covariant differentiation) that the curvature of space is supposed to manifest itself.

Again, let us take a concrete example. Suppose we are standing on the corner of Mill and University in Carbondale, Illinois, and we want to get to Oakland and Main.[25] We are not crows, so there is no direct flight from where we are to where we want to go. We must break the problem down into smaller parts, which can be solved independently of one another. So we can, if we choose, first go west down Mill, then turn north on Oakland to reach our destination. Alternatively, we can go north on University until we reach Main, and then turn west until we get to Oakland. In either case, there will be a west component and a north component in our journey, with no evident difference between a west-then-north resolution to the problem versus a north-then-west one. But what if the evidence is not all that it seems?

Suppose that after a number of times going along these different routes, we begin to notice peculiar and disorienting effects; something seems wrong, but we can not quite say what. So we set out to carefully compare the west-then-north trip (which we will just abbreviate as "wn") with the north-then-west path ("nw"). One of the tests we decide to perform is to take a compass with us as we travel along each route, carefully reading the compass at each step along the way to ensure that nothing strange happens to it, and finally recording the compass reading by marking an arrow in chalk on the sidewalk once we have arrived at Oakland and Main. We want to know if the compass reading we get at the end of the wn path – that is to say, the direction that the needle of the compass is pointing – is the same as the direction of the needle when we get to the end of the nw journey. For convenience, let us refer to the compass reading for the "wn" trip as "T_{wn}", to indicate that it is the result of Traveling down path wn, while the reading for the nw trip will be called "T_{nw}". So the question of whether these two readings are different can, in effect, be represented as

$$T_{nw} - T_{wn} \neq 0.$$

If the above is the case, then there is something about the translation of our

[25] There is nothing special in this example about the use of Carbondale. The intention here is simply to make things as definite as possible.

compass from our origin to our destination which is sensitive to the path we take.

What we have developed here, in an excruciatingly abbreviated and intuitive form, is quite analogous to the arguments for the curvature of spacetime within GR. The indices on our tensor "T" refer to the differential equations that are functioning behind the scenes; "T" itself is nothing but a book-keeping device to simplify notation. The compass in our example above stands for some appropriate vectorial quantity – i.e., something which has magnitude and, most importantly, direction, but whose other specifics we can safely ignore for the moment. Finally, the process of carefully reading the compass along the alternative paths is what is known as the "parallel transport" of the vector. Mathematically speaking, if a vector that is parallel transported along alternative paths is different at the end for the different paths, then the space around and through which the vector was transported is curved.

All of the above is merely an analogy. It is time now for the real thing. The formula below is the Riemann curvature tensor (see, e.g., Carroll 122):

$$R^{\rho}_{\sigma\mu\nu} = \partial_{\mu}\Gamma^{\rho}_{\nu\sigma} - \partial_{\nu}\Gamma^{\rho}_{\mu\sigma} + \Gamma^{\rho}_{\mu\lambda}\Gamma^{\lambda}_{\nu\sigma} - \Gamma^{\rho}_{\nu\lambda}\Gamma^{\lambda}_{\mu\sigma}$$

The Riemann curvature tensor

Now, even when looked at from a purely symbolic level, this is clearly more complicated than anything we have dealt with so far. Yet already we have all that we need to spot what is essential about this tensor in terms of its role in the claims regarding the curvature of spacetime. If we collect the first two pieces to the right of the equals sign into one group, and the last two pieces on the right side of the equation into a second group, disregarding those indices that do not change within the two groups (the rho and sigma, replaced below with dots), and finally compressing the symbolism once again for purposes of clarity, we will get something which is analogous to:

$$R^{\rho}_{\sigma\mu\nu} = (A^{\cdot}_{\cdot\mu\nu} - A^{\cdot}_{\cdot\nu\mu}) + (B^{\cdot}_{\cdot\mu\nu} - B^{\cdot}_{\cdot\nu\mu}).$$

This is just the same thing we have already seen, only doubled up and with some of the symbols compressed and dots in the place of some of the non-essential indices. And, as one might now suppose, if the Riemann tensor is zero, then spacetime is flat. It is, of course, the claim of GR that the Riemann tensor is not zero in the presence of gravitating mass, and that this shows that spacetime is curved.

However, the question which *we* must never lose sight of throughout this inquiry is just this: even if we choose to interpret a mathematical formula such as the above equation for the Riemann tensor in thoroughly geometrical terms, what does this really tell us about the world? For one thing, the claim of GR regarding the Riemann tensor is much more than that it is non-zero: according to GR, the Riemann tensor takes on a *different* non-zero value at every point throughout the universe. Spacetime, according to GR, is not only curved, but this curvature is not homogenous. As we will see, this lack of uniformity is the key to the problem we are exploring here. But is the source of this lack of uniformity in the structure of reality, or in the structure of our characterization of it?

Justified or not, such a move remains an interpretation of mathematical formulae which do not wear their meaning on their sleeve, as it were. Concepts such as "curved spacetime" are the Janus gods of physics. For, on the one hand, this is nothing more than a linguistic translation of tensorial formulae which, in their pure analytical formalism, "say" nothing at all. Mathematical equations remain, in a sense, little more than black boxes, even when their inner workings are understood perfectly. An equation is a machine that takes one collection of numbers, the input or initial values, and converts these into final values that are the solution to the problem at hand. One has an "if X then Y" type of box. But even when the "X" and the "Y" are somehow to be found in the world, properly and regularly connected not only in our theory but in our scientifically refined experiences as well, does this justify us in assuming that the structure of the world is exactly and simply the same in itself as it is in the "then" of our theory?

Eugene Wigner fretted over the "unreasonable effectiveness of mathematics in natural science" (Wigner, 222 – 37). And, indeed, there is some considerable justification at feeling astonishment over how accurate our contemporary physical theories have become in their mathematical predictions. But, is this accuracy sufficient by itself to demonstrate or even imply that there is a strong correspondence between the internal structure of our mathematical theories and the basic structures of the world? How else can we account for the fabulous successes of mathematical physics except by positing some such correspondence? In the absence of such a correspondence, how can we explain the accuracy of our theories short of an appeal to the miraculous?

However, what gets lost in the above gloss is the fact that the correspondence that is being posited – that the inner mechanism of our formal theories stands as some sort of mirror to the inner workings of the world itself – is every bit as much of a miracle as are the accurate predictions this correspondence is supposed to explain. Why should the correspondence be anything more than an imitation at the surface level brought about by a kind of "reverse engineering" of nature that is natural science? In addition, perhaps we will be inclined to a greater level of modesty on this point when we recall that these accurate theories are not Athenas which have burst fully developed from the head of any physicist Zeus. Rather, these theories are the product of the constant tinkering by entire communities of scientists on the relations between the input "X's" and the output "Y's." It is no more a miracle that General Relativity provides us with highly accurate results than it is a miracle that a modern jet flies faster, farther, higher, and with a vastly greater payload than anything that ever came out of the Wright brothers' bicycle shop.[26]

The problem of how it is possible for mathematics to be successfully applied to the world is, of course, a much more difficult problem than the above caricature might lead one to believe, and we will be revisiting this issue throughout this work. But let us note here that, despite its relevance to this discussion of the applicability of mathematics, it has seldom enough been recognized that there are significant differences between the methods of abstraction employed by the so-called pure forms of mathematical inquiry, as opposed to the applied kinds. The structures and concepts that serve as the beginning point(s) of abstraction in applied mathematics are quite different from those used in the pure areas of research, and the differences will offer possible clues to understanding how applied mathematics can accurately predict reality without directly mirroring it. In this, we will again find the work of Whitehead of considerable service.

Put briefly, it was Whitehead's argument that pure and applied mathematics start from radically different types of initial data in terms of the primitiveness of their base concepts, while their abstractive developments then proceed in more or less opposite directions.[27] For pure mathematics, logical primitiveness is the primary criterion for any starting

---

26. Once again, it is worth bearing in mind there are competing theories of cosmology which, to the limits of measurement currently available to us, are formally every bit as viable as orthodox GR.

27. Here I am largely following Murray Code in my characterization of Whitehead's ideas on mathematical abstraction (Code, chapter 7, and especially pages 174 ff). As a caveat, it should also be mentioned that by this point in Code's text, he is assuming his readers have a working familiarity with Whitehead's terminology. As such, if one simply skips to this point of his book, his argument might be very difficult to follow.

point. The more irreducibly basic a concept might be, the more acceptable it is for pure mathematics as a foundational, or "concrete" building block. The process of abstraction is then the construction of ever more elaborate syntheses of these basic units, whose validity as syntheses is ultimately demonstrated via some appropriate method of proof. These synthetic productions are, for the pure mathematician, both the exemplars and the products of high abstraction. However, for the applied mathematician, they are more akin to the applied worker's concrete starting place. More precisely, applied mathematics begins with the deliverances of experience, and then works abstractively towards what it hypothesizes as the most logically primitive elements which can be analytically distilled from the given complex. Thus, for applied mathematics, the conceptually primitive is the ultimate achievement of high abstraction.[28]

However, this process is always fallibilistic, and never capable of even a pretense of finality. For while the methods of deduction and proof employed in applied mathematics are generally every bit as robust as those used in the "pure" areas of research, there are typically no finite bounds to the number of possible deductions which could lead to some given conclusion. Since, for the applied mathematician, the "conclusions" – i.e., the deliverances of experience, which are themselves never given with utter certainty – are all that the mathematician has to start with, there is no *a priori* way to know if a particular set of abstractively derived "fundamental principles" is a "true representation" of reality, in the sense that these principles stand in some sort of direct and/or unmediated relation to equivalent factors in the world. Applied mathematicians are trying to "reverse engineer" reality. Even if, at the end of the day, they have a mathematical model that behaves as an acceptable imitation of the world it is meant to be a model of, the fact remains that all they, or we, have are sets of compatible inputs and outputs. There is nothing but our own optimism and the practical exigencies of making our theories work to assure us that the inner functionings of those theories are the same as, or even similar to, those of the world itself.

Yet, on the other hand, because mathematics is not a self-interpreting enterprise, there is also a practical necessity of assuming that – again, to some acceptable degree or other – there does exist some fairly robust

---

28. I can only mention here that this looks, at least on the surface, very much like Aristotle's distinction of that which is most knowable in itself vs. that which is most knowable by us, with the former being the starting point of pure mathematics and the end point of applied, while for the second these rolls are reversed. Whitehead does not seem to investigate this analogy himself in any of his writings.

connection between those two different constellations of inner workings. For in the absence of such an assumption, researchers in applied mathematics would have nothing with which to guide their intuitions.[29] For example, we are here to critique the assumption of a fundamental connection between geometry as a logical enterprise, and the physical realities of space so nearly universal amongst cosmologists and those working in GR. Indeed, so rare is it to find someone who tackles the mathematical machinery of GR from a purely analytical point of view, without any attempt to ground and motivate such work in geometrical intuitions, that it merits commentary verging on awe. Thus, Sean Carroll speaks of Stephen Weinberg's classic text *Gravitation and Cosmology* (1972), saying, "A great book at what it does, especially strong on astrophysics, cosmology, and experimental tests. However, it takes an unusual non-geometric approach to the material ... Weinberg is much better than most of us at cranking through impressive calculations" (Carroll, 496).

These assumptions and intuitions express themselves in conceptual forms that are not immediately mathematical in nature. As a consequence, these conceptual and intuitive notions serve as guides to mathematical thought through their analogical and metaphorical characteristics which enable mathematicians to develop constructive systems of comparison and homology which are in turn effective in the amplification of formalized symbolic structures. The notion that mathematics contains and works with analogical or even metaphorical relations is not so strange as it might first appear. Whitehead himself spoke of the geometric interpretation of tensors as "metaphors" (see above, 24). The use of analogy is even more widely recognized. George Polya devoted an entire chapter to the subject, considering its uses, its dangers, and the methodological steps involved in the precisification of initially vague analogies into sharply defined concepts (Polya 1990, 12). Edward Rothstein offered the following observation:

> Mathematics is always using analogy and proportion in its argument. In fact, when faced with a form – whether it is musical or mathematical or even in the physical world – we attempt to make analogies among the form's parts. We find in these analogies and internal mappings the

29. Though it is not a point we shall spend much time developing, it is worth mentioning here that this necessary reliance on intuition as a guide to research is not unique amongst applied mathematicians. It is an essential aspect of pure mathematics as well. For instance, see (Davis and Hersh, especially 391 ff) and (Hersh 61 ff).

fundamental proportions governing the form (Rothstein 1995, 167).

Analogy, image, intuition and metaphor all serve to inform our mathematical reasonings. Thus, the conceptual forms which shape the former group will obviously reflect themselves in turn into the latter. So even if mathematics or physics are themselves not obviously open to philosophical speculations, it is certainly the case that this former list – of analogies, images, intuitions and metaphors – is so influenced. These concepts and images, etc., in their turn certainly insinuate themselves into mathematics and physics. It is thus in this sense that we can readily understand the truth of Whitehead's claim that, "once you tamper with your basic concepts," such as space, time, etc., "philosophy is merely the marshalling of one main source of evidence, and cannot be neglected" (R 6).

Thus it is that the fundamentally philosophical character of our program once again makes itself evident. It is true that we must be able to read some mathematical symbols and formulae with a minimal degree of appreciation. But such appreciation is not the same as deductive facility, nor does it need to be. As we have seen, the basic sense of a subtle formal idea such as a tensor or a differential equation can be conveyed in a sufficiently intuitive manner that will enable us to form analogies of our own, such as the trip through downtown Carbondale, which in turn admitted of a useful abstraction, as in our pseudo-tensor "T_{wn}." This process of abstraction and analogy is itself directly relevant as an exemplar of those very same processes in physical science and mathematics. With these sorts of methodological techniques, and the variety of textual materials we will be drawing upon (which include but are scarcely limited to, Whitehead's Triad) we will be able to cultivate the needed reading skills to pursue our course through the literature of physics.

CHAPTER TWO
Chaos, Confusion and Cosmology

We shall never understand the history of exact scientific knowledge unless we examine the relation of this feeling "Now we know" to the types of learning prevalent in each epoch.

Alfred North Whitehead, "Mathematics and the Good."

I. An Outline of Relativity Theory:

This chapter will be divided into three parts. Part two will examine some of the early history of the theory of relativity, in both its special and general forms, while part three will survey the current scene of cosmology. However, this first section will be devoted to presenting an informal description of the "what" behind General Relativity: What does it claim? What does it imply? And most importantly, what is it about General Relativity that makes it an advance and improvement over the special theory? Indeed, this last question is in many respects the most important. Not only do the answers to the previous two questions emerge from the answer to this one, vital matters of what we can only characterize as the *aesthetic* dimension of science will emerge as well. Gaining some appreciation of this latter will be of considerable value to us as we encounter issues of the recent history of cosmology. A descant of some of the experimental evidence that is often cited as motivating the development of relativity will be taken up in the historical discussion of part two. Our focus here will be on the conceptual issues of relativity.

Of these conceptual issues, perhaps the most important is the idea of the "frame of reference." The idea here is simple enough: any description of the world entails a point of view from which that description is to be defined. This point of view stipulates the "zero" within the description, and establishes the perspective from which all things are to be described, and to which all things are ultimately to be referred. It establishes the frame in which the descriptive account is to be anchored. The question then arises, how are alternative accounts, given from differing frames of reference, to be related to one another? And is there a preferred frame to which all such accounts can be related?

Consider an exceedingly common example within the relativity literature, that of a moving train passing through a station, with various observers on the train and others on the platform.[30] A person on the platform, watching the train move by, will describe a passenger seen in the window of the train as traveling along with a certain velocity (which can be assigned some numerical value) in a given direction – quite obviously, the same direction and velocity as the train itself. On the other hand, that very passenger on the train can treat herself as the stationary, "zero" point of her frame of reference, and describe the world as it appears to her in equally appropriate terms. Thus, to her, the individual on the platform is the one who is moving, in the opposite direction from that which characterized the description from the platform, but with the same velocity. The question then arises, what sorts of things will be the same from these two descriptions, and what sorts of things will be different?

Well, we've already encountered one of each. For both descriptions, the numerical value of the velocity, when presented in identical units, will be the same. On the other hand, each sees the other as moving in a different direction. Complicating things a bit, suppose there is also a child on the train tossing a ball up and down.[31] To the passenger on the train, the motion of the ball will be a simple vertical one, with no horizontal displacement. However, to the individual on the platform, the ball will trace out a more complicated pattern, because there is now a horizontal component. To the person on the platform, the ball will trace an *arc* upwards and then back down, which will end abruptly when the child catches it, and then repeat when the ball is tossed up again. Nevertheless, despite these differences, both the passenger on the train and the person on the platform will describe the time taken as the ball goes up to its highest point, and then down again to the child's hand, as being the same.

These two descriptions represent two different frames of reference – one frame is referenced to the platform observer as the "zero" point, the other to the passenger on the train. Even as simple a set of examples as these brings to the fore the fact that the frame is defined not so much as the absolute *position* of any particular observer who might be occupying some individually stipulated "zero point," but rather by the relative motion(s) of the world in respect to that which has been chosen to be treated as at rest.

---

30. For example, see (Einstein 1961, pp. 15 ff). When we expand the example from trains to any moving vehicles, which is the essential point of the example, i.e., comparing the alternative frames of reference produced by differently moving observers, then the exemplars are as common as the accounts of relativity itself. One can even find an example of railroad cars and rockets in Earth orbit (Taylor and Wheeler 1963, pp. 8 ff).

31. Again, an extremely common refinement of the train example.

Whatever is at rest *relative* to some selected X – defined as that relative "zero point" – is in the same frame as that X, and in that respect "at rest." In our examples above, we have two frames defined by the positions of observers who see themselves at rest – whether as a passenger on the train, or the individual waiting on the platform – with respect to the rest of the world. Now, the mathematics relating the two frames of the observers on the platform and on the train to one another is quite straight-forward and well understood, and as such we may safely ignore the technical details. What we want to focus on instead is the question, what happens when the objects of our interest in various frames are no longer mundane objects like balls tossed by children, traveling at the comparatively pedestrian speeds of even the fastest trains? In particular, what happens when we bring electromagnetic phenomena, most especially light, into the picture?

When we do this, the situation immediately becomes more complicated. For while we were only dealing with things like a child's ball, we remained within the purview of simple dynamical systems that could be adequately described by the mathematics of Newtonian mechanics. Now, however, we must bring to bear that astonishing achievement of 19th century science, the mathematical unification of electromagnetic phenomena produced by James Clerk Maxwell. Justifiably characterized as the "the most spectacular (mathematical) triumph of the nineteenth century," (Kline, vol. 2, pg. 698) Maxwell's equations are quite arguably the single most important scientific development immediately preceding, and leading up to, that of relativity theory itself.[32] The equations, first derived by Maxwell in 1864 then published in 1865, ushered in a whole new era of physics the implications of which are still being worked out.

What changes with Maxwell's equations are the rules of transformation which make it possible to compare the descriptions produced within one frame of reference to those developed in another. In our earlier example of the train, these rules were quite simple developments of ideas from Galilean and Newtonian mechanics. With Maxwell's equations the difficulties that emerged were not so much due to the complexities of the equations themselves – although, it should be noted, these formulae "raised the bar" on the subtleties of use of partial differential equations beyond anything that had come before them. Rather,

---

32. This specifically includes the famous Michelson-Morley experiment. As we shall see below, the role of this experiment in the development of Einstein's ideas is open to question, whereas that of Maxwell's equations is not. Indeed, in the absence of Maxwell's work, it is scarcely conceivable that the theory of relativity would have been even an imaginative possibility.

what makes the new situation so much more difficult to manage is the need to develop intelligible rules for the translation of descriptions of electromagnetic phenomena from one frame of reference to another in such a way that the rules of Maxwell's equations are not violated. How is one to keep the rules *constant* within differing frames of reference, when those different frames are *moving* with respect to one another?

The ideal of physics, of course, is not simply to generate an individual description of an arbitrary degree of mathematical precision for each and every possible frame of reference. If this were the case, any *ad hoc* system of formulae would be more than adequate for each and every situation encountered. They could then be individually manipulated, extended, and fine tuned to taste, without any regard for their relations to any sort of comprehensive system of relations. In particular, there would be no need to relate the descriptions of one frame of reference to another. But such a relational system is precisely what most motivates physicists. Ad hoc, iconoclastic and unsystematic characterizations of phenomena in their uniqueness, however formal and precise they might otherwise be, serve no more than the most temporary and locally instrumental functions in science. Thus, for instance, in our initial train example, there was an underlying reality that was taken to be common to both observers, whether on the platform or on the train. This common reality found itself represented in the underlying *uniformities* expressed by the rules of transformation which made possible the comparison of the two frames of reference. Different descriptions from differing frames might present alternative quantitative relations for what otherwise appear to be "the same" things. But these differences are themselves an expression of the deeper *identity* of the underlying rules for generating these quantities. "Physical quantities differ in *value* between the two frames but fulfill identical *laws*" (Taylor and Wheeler, pg. 13, original emphasis). In the older view of physics, these laws could, in turn, be traced back to yet more fundamental principles, until at last one reached the ultimate frame of reference in the form of Newtonian absolute space and time. This was, in essence, a "God's eye view" on the universe, a point of view to which all other points of view could be referred. And, indeed, within this theoretical setting, *had* to be referred in order to have any meaning.

Now, Maxwell's equations represent a set of laws that are – that *must* be – the same between all different frames, including the supposedly absolute space and time of the Newtonian scheme. But discovering the rules of transformation that would keep the underlying laws of electromagnetism constant between different frames of reference proved enormously more difficult than the preceding dynamical case of the train. No small part of the difficulty in resolving this problem was that older Newtonian scheme which had, up until the mid-nineteenth century, proven so successful, could not be given any pride of place with Maxwell's equations. Relating other forms of dynamical systems to an ultimate frame – the Newtonian absolute space and time – worked so well that scientists had almost as much trouble divorcing themselves from that notion as they had trying to shoe-horn Maxwell's laws of electromagnetism into it.

One of the chief obstacles to such a melding was the profound lack of symmetry between space and time conceived as absolute vessels, and other frames of reference. There was a fundamental imbalance in the nature and direction of the relations from the one to the other. While "ordinary" reference frames stood on something of an equal footing in terms of how they related to one another (no one of them was privileged as being ultimate), the Newtonian absolute frame, to which all others were in a sense secondary, held no such relative position. This lack of relativity on its part created a variety of problems, some nearly and others simply insoluble, in the context of electromagnetic theory. It is at this point that Special Relativity comes into the picture.

Special Relativity dethroned absolute space and time, and presented a picture of the physical world as something that was more like a democracy of frames of reference. It not only continued to be the case that the laws *within* each frame were identical (as stated in the quote above from Taylor and Wheeler), this identity became, in a sense, its own *raison d'etre*. It was no longer viewed as necessary to refer these laws back to some ultimate and absolute frame of reference in order to give them meaning. Rather, the laws of physics came to be seen as sufficient unto themselves. It is worth noting in this regard that the quote above from Taylor and Wheeler, while genuinely applicable to all areas of physics, was in fact originally made specifically regarding Special Relativity. By dropping the *unique* frame of absolute space and time, physics could finally treat *all* frames of reference as being on the same footing, thereby making it possible to derive a thoroughgoing identity of physical laws as operating within and between them. This identity manifested itself in the new rules for translating descriptions between frames, the Lorentz transformations. Named for the Dutch mathematical physicist Henrik Antoon Lorentz, this system of transformations was the first successful technique developed for relating alternative frames of reference to one another in such a way that the Maxwellian laws of electromagnetism remained identical throughout. These rules succeeded because they were able to find and express a higher degree of *symmetry* than had previous physical theories.

So, what exactly is meant by "symmetry" in the above? Intuitively, the notion of symmetry involves a kind of sameness amongst the parts of a thing. We think of most animals as expressing a kind of symmetry (specifically, *bilateral* symmetry) because the left side of an animal (such as a human being) is very similar to the right side. This kind of similarity is not an identity – the left and right sides are more like mirror images of one another, and even then only partially so – but the relation is certainly robust. When we turn to artificial structures such as various geometrical shapes, the number and quality of the symmetries we can find becomes even more striking. Thus, for example, if we draw a line along the diameter of a circle, the two sides are now entirely the same. But a circle has more than one line splitting it in half. Indeed, there is no finite bound to the number of lines[33] which divide the circle into fully symmetrical halves. Moreover, we can do more than just compare the different halves of a circle. We can also rotate a circle, to the left or right, without fundamentally changing either the circle or its relations to the space it is in (such as the page on which a circle is printed), or to other geometrical forms which might be near it, etc. Thus, we find that the circle has "rotational symmetry" as well. There are a variety other kinds of symmetry, but what they all share in common is this idea of some sort of specifiable form of sameness within the context of some complementary manner of operational change, such as mirror reflections, rotations, etc. Or, in somewhat more technical language, what identifies a mode of symmetry is the "invariance of a configuration of elements under a group of automorphic transformations" (Weyl 1980, preface).

This is an important definition, and merits our close attention. "Invariance of a configuration" simply means that a shape, form, or structure is not changed by some action. The circle is still a circle, even after it has been rotated. An "automorphic transformation" is an action "which preserves the structure of space," i.e., one in which however much things are changed around, the rules of geometry are not touched. This should immediately remind us again of the above quote from Taylor and Wheeler – i.e., values might change, but the laws remain identical. Finally, the automorphic transformations are systematically related, we are told, as a "group," a concept which must certainly be characterized as one of the

---

33. The incautious way of expressing this is to say that there is an "infinite number" of such lines of symmetry.

most important in all of mathematics and physical science.[34]

The set of rules as to what constitutes a group seem, on the surface at least, so patently simple that one can scarcely be blamed for wondering how the whole concept fails to collapse into triviality. A group is just a set of "elements" with a rule of composition of those elements such that the composition of any two elements of the group is the same as (or picks out) another element of the group. In addition, there is a distinguished element (or, alternatively, an element distinguished by the rule of composition), such that the element picked out by the composition of the distinguished element with any other element is the same as that other element. This distinguished element is also referred to as the "identity" element of the group. Finally, every composition is reversible, which is the same as saying that every element has an inverse under the group's rule of composition. More formally, if we take "G" as our group, "a," "b" and "c" as arbitrary elements, and represent the composition of "a" with "b" as "ab," we can express these relations formally as:

(1) $\forall(a,b \in G), \exists(c \in G)\ (ab = c)$;	i.e., groups are closed under composition,
(2) $\exists(e \in G), \forall(a \in G)\ (ae = a = ea)$;	i.e., there is a distinguished, identity element,
(3) $\forall(a \in G), \exists(a' \in G), (aa' = a'a = e)$;	i.e., every element has an inverse.

Now, groups can be studied in their own right, as abstract mathematical objects. But they can also be viewed as "concrete" entities[35] where the "elements" are not just abstract "things," but specific operations which act on some other structure or object. These operations could be

34. Detailed and comprehensible treatments of group theory can be found in any textbook on abstract algebra, such as (Birkhoff and MacLane 1997). Of course, (Weyl 1980) is entirely devoted to the connections between symmetry and group theory. The centrality of the idea of a mathematical group becomes more readily apparent in such historical treatments as (Kline 1972), or (Kramer 1980). For instance, Edna Kramer says, "The group concept is of inestimable importance in modern mathematics and physics, so much so that the great Henri Poincaré ... once brashly said, 'the theory of groups is *all* of mathematics.' At any rate, we shall be alluding to groups in so many later chapters that we shall not give specific references at this point" (Kramer, pg. 81, original emphasis). The implication of this last being that, while the formal relations of groups can be glossed in a single chapter, the historical relations cannot be so neatly contained.

35. Well, concrete to a mathematician.

such things as rotations, the taking of mirror images, etc., while those other structures or objects which are operated on by the elements of the group could be geometrical forms, physical objects, and so on. This, of course, is the tie-in between group theory and symmetry. Those group actions which leave the underlying structure "the same" reveal a kind of formal symmetry in that object. For instance, picking up an otherwise totally irregular rock and moving it a meter to the left is a kind of symmetry of the rock (namely "translational" symmetry), since the rock can be moved back (the action has an inverse) or it could have been left alone to begin with (the translational movements have an "identity" element). Another thing to notice is that it is *not* in general the case that the composition ab = ba for any arbitrary group G. Consider for instance, the group composed of the rotations of the faces of a Rubik's cube. If we define "r" to mean the rotation of the face with the red center ninety degrees to the right, and "b" the same for the adjacent blue centered face, then an action which first rotates the red then the blue, does not have the same effect as blue then red. I.e., $rb \neq br$.[36]

In any case, despite their evident simplicity, it turns out that so vital are group theoretical relations in characterizing the invariant structures of any particular geometrical space, i.e., determining its symmetries, that Felix Klein recognized that different kinds of geometries could be identified by the group theoretical invariances specific to them. Klein used this idea to launch his famous "Erlanger Program" which aimed at the complete unification and characterization of all geometries (Kramer, chapter 17, for the details).

There is, however, a great deal more to the idea of symmetry in mathematics than just its formal character. Symmetry itself is very much an *aesthetic* aspect of a formal theory, and not just a mathematical one. That there is an intimate connection between symmetry, beauty, and mathematics is scarcely a recent discovery. We find, for instance, in Aristotle, the observation that, "those who assert that the mathematical sciences say nothing of the beautiful or the good are in error. ...The chief forms of beauty are order and symmetry and definiteness, which the mathematical sciences demonstrate in a special degree" (Metaphysics, 1078a 33 – 1078b 2, quoted from McKeon 1941). Furthermore, since symmetry comes in many kinds and flavors, the choice of which forms are

---

36. It should come as no surprise that there are entire books devoted to this subject. See, for instance, *Adventures in Group Theory: Rubik's Cube, Merlin's Machine, and Other Mathematical Toys*, by David Joyner, Johns Hopkins University Press, 2002.

"essential" in any particular inquiry cannot be driven by purely logical considerations. While it is surely the case that in any physical science empirical matters will play a central role in any such decision, even these will occasionally not suffice to determine the matter. As we shall see, in this chapter and throughout this investigation, a measure of aesthetic "taste" also plays a part.

This gives us enough information to return to the Lorentz transformations. The Lorentz transformations are a mathematical group, which means they pick out a particular geometrical space as the invariant structure which forms the basis for their operations (Kramer, 434). This is the space of Special Relativity, which is formally quite sound. Indeed, mathematically speaking, it is quite a bit more satisfying than that of Newtonian physics, because of its superior levels of symmetry. But intuitively and aesthetically it can be rather disconcerting. Things happen in this space that just seem odd. For one thing, the whole notion of rigid bodies, which retain their physical dimensions in all frames of reference, appears to have been sacrificed. The kind of symmetry in which rigid bodies retain their rigidity in all frames of reference has been relinquished in order to retain throughout all frames the deeper, but more abstract, symmetry of the identity of the laws of electromagnetism. This leads to the (in)famous "Lorentz contraction"[37] where the dimensions of a nominally rigid body shrink in the direction in which it is moving to a degree which is proportional to the ratio of its velocity to that of the speed of light[38] (Taylor and Wheeler, 39 – 47 and 80).

Another puzzle for early interpreters of the Lorentz transformations was the effect known as the dilation of time, a temporal contraction similar on a formal plane to the contraction of physical objects in space. The most famous illustration of this is the so-called "twin paradox." Two twins begin in the same place, a non-accelerated frame of reference. But one of them boards a space ship which travels at very high velocity to some distant point in space, then turns around and returns to the first twin who has remained in the comparatively stationary reference frame. When the door of the space ship opens, the twins are confronted by the peculiar fact that the twin that rode the space ship is significantly younger than the one that remained behind. Somehow, time has not passed at the same rate for the

---

37. Also sometimes referred to as the "Lorentz-Fitzgerald contraction," since G. F. Fitzgerald had independently hypothesized a similar idea (Holton, pg. 301).

38. Specifically, the length of the body shrinks in the direction of travel as a multiple of $(1 - v^2/c^2)^{1/2}$, where "v" is the velocity of the "rigid" body, and "c" is the speed of light in a vacuum.

two, evidently treating the traveling twin much more kindly.

Both of these puzzling aspects of special relativity are resolved (mathematically) by anchoring that theory within a Minkowskian *four-dimensional spacetime* manifold. Rigid bodies remain rigid, but they appear to have contracted when moving because what we observe is not the complete four-dimensional object, but only its moving projection into the three-dimensional space or our ordinary experience. Again, the twin "paradox" only seems strange to us because all we see is the three-dimensional shadow cast by events that are four-dimensional in "reality" (at least, according to the theory) (Carroll, 10 – 11).

But there remains a vital and galling asymmetry within special relativity. This theory is restricted in the sense that its range of application is predicated around what are known as "inertial" frames of reference, frames which, if they move at all, move without any acceleration. In particular, since an inertial frame is unaccelerated, that means it is also free from the influences of gravity. So, for example, what does it mean in the twin paradox that one twin remains "comparatively stationary"? But "stationary" relative to what? Evidently the "stationary" twin is entirely outside of the reach of gravity, since this means outside of any accelerating influences. But in comparison to what is that twin not moving? It would appear that such non-accelerated frames are analogous to the absolute space of Newtonian physics, in the sense that they hold a privileged position within the special theory, a position that is not fully explained. As we have seen, this lack of symmetry in how different frames are situated in a theory is regarded as an imperfection, an aesthetic blemish to be overcome if at all possible. However, it turns out that when the effects of gravity are brought back into the picture, it becomes possible to *generalize* the structures of special relativity into a more symmetrical system, the system of GR.

But gravity is not something that is just added on to special relativity. Gravity in general relativity is interpreted as the variable *geometrical* structure of four-dimensional spacetime. In this respect, gravity becomes the paradigm of the accelerated vs. the unaccelerated frame of reference. As such, gravity in GR gives us the missing information to say what an inertial frame of reference "really" is. To whit, an inertial frame is that frame of reference where everything is moving in such a way that it is offering no resistance to the natural push and pull of all the surrounding gravitational forces. Thus, for example, in the above story of the twins, even if the twin on the spaceship did not realize she was on a ship that was expending energy to move her away from her sister, she nevertheless could tell that she was the one accelerating away from her twin, and not the other way around, by observing how her movements were cutting across and through the natural distributions of gravitational forces. But GR achieves this higher level of symmetry by erasing a distinction which, initially, seems like a very appropriate distinction to make. Space and gravity are no longer two fundamentally different kinds of phenomena or modes of relatedness. It is almost impossible to overstate how closely related space and gravity are within general relativity. It is not the case, in GR, that we are dealing with space *and* gravity. Rather, what we are dealing with might better be called space *as* gravity.

Let us look more closely at this. Now, to begin with, general relativity is a "classical field theory," which is to say that the mathematical characterization of GR places it more on a par with such theories as Newtonian physics or Maxwell's electrodynamics than with that other major achievement of 20th century physics, quantum mechanics. The "field" is most essentially manifested by what is known as the "metrical tensor," a mathematical structure which Einstein (and others following him) called the "fundamental tensor" (Einstein *et al*, 127 ff). This special tensor is a function defined at every point in spacetime, specifying the mathematical representation of the basic physical and metrical characteristics of the universe at that point. The space of these points, which is all of space itself, forms a "tensor field" which can then be operated on by other functions,[39] in turn further developing the relational structures of the geometry of spacetime. Some of these other operators – for example, the "Lagrangian," whose technical details we need not fret over here – are extremely useful not only in developing the underlying relations and symmetries of GR, but they also provide a concrete basis for developing the connections and homologies between GR and, say, Newtonian mechanics (Carroll, 37 ff). It was, of course, in the context of this latter theory that the eponymous Lagrange originally developed his operator,[40] and its appearance in GR is a source of useful associations. Such connections can provide valuable mathematical insights into a theory in terms of the formal analogies developed by a common operator in alternative systems. It can also provide valuable, even if informal, heuristic guides to both the intuitive and aesthetic dimensions of a new theory, when a well developed structure (such as the Lagrangian) finds applications in a

---

39. A function which operates on another function, or field of same, is often called a "functional."

40. See, for example, (Kline) or (Kramer), throughout, or any volume on the history of mathematics or physics.

newer context.

However, we must not permit these comfortable homologies to blind us to the radical differences between GR and classical physics. There is, in particular, one essential difference between GR and either of Newtonian mechanics or Maxwell's theory of electromagnetism. Because the *metric* tensor is the basis of the field in GR, there is no separation between fundamental measurable – i.e., metrical – quantities at a point, and the physical influences acting at that point. The geometrical and the physical no longer stand as separable structures. As a result the concept of "straight line," and consequently of distance itself, has no independent status within general relativity. These things are variable, and their variations are in part due to the contingent physical factors which influence the metrical tensor at any given point. In contrast, if we look at Newtonian cosmology, these metrical structures are built into the *geometry*, but this geometry is independent of the physics. As such, physical events cannot alter metrical relations within the Newtonian system.

Not so in GR. The metrical tensor is not a constant which expresses the same values and relations at every point in space. Rather, it is a *function* of each point in spacetime which is itself operated on by other function*als*. Thus, what constitutes the "straightest line" between two points in GR (also known as a "geodesic") cannot be described in a manner that is independent of the physical, material structures of reality which give the points along that geodesic their particular metrical characteristics. Moreover, it turns out that this "straightest line," this geodesic, is precisely that path through spacetime which determines the *inertial frames* of reference. So, according to GR, objects within inertial frames – which is to say, objects that are unaccelerated, which in turn means that they are following the contours of gravitational effects without any resistance to them, or any other external contributions to their motions – are traveling in the straightest lines possible in nature. This tightly bound whole, of gravity determining the geodesic structure of space, which in turn defines inertial motions, is what couples the geometry of GR directly to the physics. So even though it is true, on the one hand that, "general relativity itself is a classical field theory," one must then add that, "It is nevertheless fair to think of GR as somehow different; for the most part other classical field theories rely on the existence of a *pre-existing spacetime geometry*, whereas in GR the geometry is determined by the equations of motion" (Carroll, 45, my emphasis).

One might add here that, precisely because the geometry of GR has no footing independent of the “purely”[41] physical relations, it becomes prohibitively difficult to conceptualize GR in its full generality, with all of its symmetries readily evident and in a form that allows for any sort of high level manipulation, if everything in the theory is presented in a “coordinate form.” By “coordinate form” I mean, of course, something similar in kind to the Cartesian geometry we've all studied in our youth, where forms and functions are laid out in a well defined, multi-dimensional grid (such as a two-dimensional piece of graph paper) with each point labeled by its “x” and “y” coordinates. Coordinate systems are inherently wedded to some particular frame of reference by their very need to have a “zero point” defined, to which all the other coordinates can be referred. Such coordinate driven methods certainly have a role in GR, especially when the need has arisen to produce actual, concrete numbers regarding some particular prediction or observation. But for the purposes of general theoretical development, a more general scheme of representing things is essential to the process. This is certainly one of the most important reasons for the use of tensors in this theory and other theories of spacetime and gravity. For tensors, interpreted as geometrical objects, are independent of coordinates and automatically encode the essential invariances – i.e., the symmetries – of the theory, without having to refer that theory to any specific frame.

This, finally, is what sets our problem for us. For we will find that the identification of the geometry within GR with the physical relations of gravity undercuts the possibility of a coherent theory of measurement within that theory. This identification is most profoundly evident in the geometrical interpretation of the tensors, those “geometrical metaphors” to which Whitehead objected, and for just this very reason. So much, then, for our gloss of the ideas of general relativity. It is time now to look at some of the empirical and historical background from which those ideas emerged.

II. Cosmology before 1960:

While there are a variety of possible ways of organizing this subject matter, for our purposes cosmology in the 20th and early 21st centuries naturally divides itself into three periods. The first three decades of the 20th century were a time of considerable confusion and dispute as the older, Newtonian cosmology fell further and further out of favor.[42] This was the

---

41. Obviously we must be careful here, since what is or is not “purely” physical is going to depend very much upon the theoretical framework from which one is making such a claim.

42. My treatment here is necessarily very brief and extremely orthodox in terms of the ideas presented. I shall have nothing to say about David Hilbert's publication of a

period of time when Einstein's theories emerged as the new consensus regarding the macro-structure of the universe. The second period is the next thirty years or so, ending in the late fifties or early sixties. This was an era when relativity reigned largely unchallenged in cosmological circles, but during which time there was notably little empirical pressure on GR due to the lack of sufficiently refined technologies capable of seriously testing its predictions. The third period, from the early 60's onward, was a growth period for cosmology as techniques emerged that could finally begin to push the boundaries of GR with some earnestness. This period is also marked by the growth of a kind of "cottage industry" of alternative theory formation, as a series of new theories of gravity were proposed as potential competitors of GR. Finally, let me just add that later analysts may see a *fourth* period as having begun in the mid nineteen-nineties, which is when the data from the Hubble space telescope began to throw all cosmological theories to the lions with a stream of observations that has proven exceptionally difficult to interpret. However, we will not touch on any of the details of these findings until the final chapter.

Our story begins with a different set of empirical difficulties facing the then dominant theory of space and time. There is the oft repeated account of how Einstein's theory came about as a response to a crisis in cosmology at the turn of the twentieth century, driven primarily by the experimental evidence of the famous Michelson-Morley experiment. This experiment was the attempt to directly measure the effect of the movement of the earth through the "luminiferous ether"[43] on the speed of light, by comparing the travel times of two light beams moving perpendicularly to one another over a physically identical distance, with one such beam pointed in the direction of the earth's movement through space. Now, in the 19th century, the only physical process capable of measuring light with any sort of accuracy was light itself. The possibility of using light as its own yardstick in the attempts to get a direct handle on the ether was made a reality by Albert Abraham Michelson's invention of the interferometer in

relativistic theory of gravity a mere five days before Einstein's famous paper, nor of Poincaré's important contributions to the theory of measurement and time. See the citations in the References under (Poincaré) and (Hilbert). Whitehead was certainly aware of Poincaré's early work on science and convention, which he explicitly mentions in (CN, pp. 121– 3). Also, I will avoid without prejudice discussing things in terms of Kuhnian "scientific revolutions," though there are clearly aspects of this account which lend themselves to such a reading.

43. The theoretical medium which was thought to be the necessary medium of the wave aspect of light, in much the same way as sound waves need some kind of material medium such as air or water.

1881. Using this instrument, Michelson, along with E. W. Morley, attempted to measure infinitesimal changes in the speed of light due to the movement of the earth through the ether.[44] But by 1887, their efforts had entirely failed. It was well known that Michelson's interferometer was more than adequate to the task of measuring such effects, even well below the level of those predicted by the then current theories, so this null result of the Michelson-Morley experiment led to a crisis in theoretical physics. H. A. Lorentz struggled valiantly to come up with an explanation for the null results, but it was only with the 1905 publication of Einstein's "On the Electrodynamics of Moving Bodies," which heralded the overthrow of Newtonian mechanics and the introduction of special relativity, that the null results of the Michelson-Morley experiment came to be fully accounted for.

This is a comfortable story that fulfills our basic expectations of how science is supposed to operate: the crucial experiment, which leads to the scientific crisis, which in turn is finally resolved by the radical new theory. It is a widely accepted narrative,[45] which has achieved something of the status of a cultural icon of science. The problem with this story, though, is that it is at best a considerable oversimplification; at worst, it is quite possibly an outright fairy tale.

On the face of it, challenging this popular story of the emergence of relativity seems scarcely reasonable. For instance, according to Adolf Grünbaum, to question this general notion of the development of relativity theory, "leaves us ... puzzled concerning the logical, as distinct from psychological grounds which would then originally have motivated Einstein to have confidence in the principle of relativity without the partial support of the Michelson-Morley experiment, while that very lack of support would have sufficed, by his own admission, to assure the abandonment of the principle 'without qualms' by his colleagues" (Grünbaum, 381). As Grünbaum goes on to point out, Einstein himself on a variety of occasions acknowledged the experiment as a critical element in the development of his own ideas, and their ultimate vindication (*ibid*). Indeed, Grünbaum cites Einstein's own mention of "the unsuccessful attempts to discover any motion of the earth relatively to the 'light medium'

---

44. Descriptions of the details of this well known experiment, as well as the operational structure of Michelson's interferometer, are extremely common in both the popular and technical scientific literature, and need not concern us here. See, for example, (Taylor and Wheeler, pp. 76 – 78).

45. Gerald Holton lists five popular sources, including the Feynman Lectures on Physics, and his own work from 1952 (Holton 1980, pg. 270 and fn. 25)

[aether]," and demands that the critics of this connection,[46] between Einstein and the Michelson-Morley experiment, explain why such an evidently explicit reference to the experiment in Einstein's 1905 paper does not simply nail the case for the connection (Grünbaum, 380).

One of the claims Grünbaum believes does not reconcile with what he views as the established facts is Gerald Holton's assertion that Einstein's 1905 paper "begins with the statement of formal asymmetries or other incongruities of a predominantly aesthetic nature (rather than, for example, a puzzle posed by unexplained experimental facts)" (Grünbaum, 380; Holton, 168). Grünbaum's demand is rather difficult to interpret charitably, though, given that Holton's statement is so clearly and simply true. Consider the opening three sentences of Einstein's 1905 paper:

> It is known that Maxwell's electrodynamics – as usually understood at the present time – when applied to moving bodies, leads to *asymmetries* which do not appear to be inherent in the phenomena. Take, for example, the reciprocal electrodynamic action of a magnet and a conductor. The observable phenomenon here depends only on the relative motion of the conductor and the magnet, whereas the customary view draws a sharp distinction between the two cases in which either the one or the other of these bodies is in motion (Einstein *et al*, 37; my emphasis).

Einstein adds another three sentences to this opening paragraph, providing some further detail as to how actual magnets and conductors behave when one is moved in the vicinity of the other. Einstein opens the second paragraph of his 1905 essay with the following sentence, already partially quoted above, but presented here in full:

> Examples of this sort, together with the unsuccessful attempts to discover any motion of the earth relative to the "light medium," suggest that the phenomena of electrodynamics as well as of mechanics possess no properties corresponding to the idea of absolute rest. (*ibid.*)

Holton raises two points in particular regarding this sentence: first, if Einstein is so simply inspired and motivated by the Michelson-Morley experiment, why does not he ever mention it by name? Second, and far more tellingly, Einstein's comment on the "unsuccessful attempts" is further ambiguated by the fact that he could have had "any two or more of

---

46. In this case, Gerald Holton, whose replies appear in (Holton 1980), especially chapter 9. The paper which Grünbaum is criticizing appears as chapter 5 in (Holton).

at least seven experiments" in mind (Holton, 302). Indeed, there is no necessary reason to suppose that Einstein had *any* specific experiments in mind: his remark, as it stands refers to nothing more than a *generic* understanding of the then current status of experimental efforts.

Furthermore, Einstein's own testimony on the subject is inconsistent at best. Grünbaum is correct in citing places where Einstein explicitly names the Michelson-Morley experiment as directly relevant to the development of his own theory. But there are other places, easily as numerous and reliable as the sources Grünbaum selects, in which Einstein equally explicitly denies that the experiment had any significant role in his thought (Holton, chapter 9, throughout). In yet other places, Einstein does discuss the experiment and the puzzling results it gave in light of the theory of the day. But he then goes on to say that the Lorentz-Fitzgerald hypothesis adequately accounted for those facts (Einstein, 58), or that Lorentz successfully demonstrated how "the result obtained at least does not contradict the theory of an æther at rest" (Einstein, 168).

While some issues were eventually discovered with Lorentz's solutions (Holton, 303) at the time there was thought to be no empirical basis for preferring Einstein's relativity theory over Lorentz's contraction hypothesis. Thus, for example, Ernst Cassirer, writing in the early 1920's, observed that

> As is known, the investigation of Michelson and Morley ... was explained as early as the year 1904 by Lorentz in a manner which fulfilled all purely physical demands. ... An experimental decision between Lorentz's and Einstein's theories was thus not possible; it was seen that between them there could fundamentally be no *experimentum crucis*. The advocates of the new doctrine accordingly had to appeal – an unusual spectacle in the history of physics – to general philosophical grounds.... (Cassirer, 375, original emphasis.)

There are several points worth noting here. First, Cassirer's essay had the benefit of being read and commented on in manuscript form by Einstein himself, and as such stands as a particularly valuable testament on the then contemporary situation in physics (Cassirer, 349). Secondly, Einstein expressly denies having any awareness of Lorentz's 1904 paper, mentioned above by Cassirer and included in (Einstein *et al*), in which Lorentz's solution to the problem posed by the Michelson-Morley experiment is offered, prior to the publication of his own 1905 essay (Einstein *et al*, 38). Consequently, while Lorentz was certainly motivated by the Michelson-Morley experiment, one cannot draw a similar line of influence toward

Einstein's work. On the other hand, the "general philosophical grounds" Cassirer mentions are, among other things, the higher levels of *symmetry* found in Einstein's theory. "[F]rom the standpoint of the theory of relativity [Lorentz's] solution of the difficulty was the right one. But on the basis of the theory of relativity the method of interpretation is incomparably more satisfactory. According to this theory there is no such thing as a 'specially favored' (unique) co-ordinate system" (Einstein, pg. 59). Meanwhile, even amongst contemporaries who believed strongly in a direct connection between the Michelson-Morley experiment and Einstein's theory, the evidence occupied an ambivalent position. Reichenbach, writing in 1928, acknowledged that relativity "does not follow, of course, from this experiment alone" (Reichenbach, 195, footnote). On the very next page, Reichenbach goes on to say that, "The relativity of simultaneity has nothing to do with the contraction in Michelson's experiment, and Einstein's theory explains the experiment as little as does that of Lorentz" (Reichenbach, 196).

Among other contemporaries of Einstein, the connection between the Michelson-Morley experiment was even less evident. For his part, Michelson long continued to view his experiment as a *failure*, and looked upon Einstein's theory with considered displeasure (Holton, 317 ff). Nothing about his 1887 work with Morley is even mentioned in any part of Michelson's 1907 Nobel prize in physics. The prize itself was given for Michelson's development of "optical precision instruments and the spectroscopic and metrological investigations carried out with their aid." Nowhere in either the award speech given by K.B. Hasselberg, or in Michelson's Nobel lecture, is the word "ether" to be found.[47] Holton points out that wide spread acceptance of Einstein's theory even within the German physics community does not appear to have occurred prior to Minkowski's 1909 paper (found in Einstein *et al*, 73 – 91), in which he embedded Einstein's special relativity within the geometrical structure of four-dimensional spacetime (Holton, 268 – 9).

Well into the nineteen-twenties, when Einstein's theory was attaining the status of the dominant cosmological theory in physics, it was still necessary for philosophical treatments of the subject to argue against the notion that there are *a priori* reasons for taking Euclidean geometry as the "true" logical form of space. Thus, for example, Ernst Cassirer, arguing from within a neo-Kantian tradition, devoted an entire chapter to the subject, rejecting any priority to the rule of Euclidean axioms (Cassirer,

---

47. These facts and others, including the full text of both speeches, can be found at the official Nobel organization's website http://www.nobel.se/physics/laureates/1907/.

430 – 444). Or, again, Reichenbach, writing only a few years after Cassirer, and adamantly opposed to any form of Kantian or neo-Kantian philosophy, argued emphatically for the "relativity of geometry," and against any sort of *a priorism* (Reichenbach, 30 ff., and throughout). Almost twenty years after Minkowski framed special relativity within a four-dimensional spacetime manifold, and ten years after Eddington's triumphal return from the Indian Ocean expedition, proclaiming the success of Einstein's general relativity with its Riemannian geometry, it was still considered necessary to argue against giving pride of place to Euclidean geometry as representing the true structure of space. It is not merely as a matter of historical curiosity that these arguments are mentioned here. Rather, these arguments form the context in which Whitehead advanced his own theory, a context which undoubtedly contributed to the systematic misunderstanding of that theory.

Whitehead was no advocate of the kinds of *a priorism* that were critiqued by the above authors, stating that "the character of time and space is not in any sense *a priori*" (R 64). However, Whitehead was also adamant that this did not mean that the structure of space and time was entirely a matter of empirical inquiry. There is, for Whitehead, an "essential relatedness of any perceived field of events to all other events," which does not admit of just any sort of connection. These relations, Whitehead goes on to insist, "must possess a systematic uniformity in order that we may know of nature as extending beyond isolated cases subjected to the direct examination of individual perception." (*ibid.*) Far beyond any particular mathematical formulae presented in the later sections of his work, this principle of uniformity is the single most important feature of Whitehead's theory. It is this principle which led Whitehead to "the rejection of Einstein's interpretation of his formulae, as expressing a casual heterogeneity of spatio-temporal warping, dependent upon contingent [circumstances]"[48] (R 65).

And, as we have seen above, this is precisely the case in Einstein's theory. Because the distinction between geometry and physics has been eliminated within GR, the very shape of space at any given point, as this is expressed by the metric tensor, is functionally dependent upon the gravitational influences at that point which, in turn, determine the values of the metric tensor. Since these gravitational influences are contingent and non-homogenous, so is the geometry of space. For, "if space-time be a

48. The word Whitehead actually uses at the close of this sentence is "adjectives." This is a technical term within Whitehead's text, the explanation of which at this time would take us too far afield.

relatedness between objects, it shares in the contingency of objects, and may be expected to acquire a heterogeneity from the contingent character of objects" (R 58). Whitehead argued that this lack of uniformity led to catastrophic problems with Einstein's theory:

> I cannot understand what meaning can be assigned to the distance of the sun from Sirius if the very nature of space depends upon casual intervening objects which we know nothing about. Unless we start with some knowledge of a systematically related structure of space-time we are dependent upon the contingent relations of bodies which we have not examined and cannot prejudge (R 59).

In other words, in the absence of an established set of rules that allow us to understand the metrical relations which contribute to the resultant value of any given measurement, we have no way of knowing *what* that measurement is telling us. Consequently, Whitehead saw the necessity of developing an alternative theory of relativity (the *applied* portion of R) which was built around a system of uniform geometrical relations, such as would allow a meaningful theory of measurement to be developed. "The structure is uniform because of the necessity for knowledge that there be a system of uniform relatedness in terms of which the contingent relations of natural factors can be expressed. Otherwise, we can know nothing until we know everything" (R 29 – 30).

The system of uniform relations that Whitehead chose to employ in his own applied theory was that of Euclidean geometry. This was not an arbitrary decision, but neither was it an absolute or *a priori* one. We have already seen above that Whitehead rejected any *a priori* preference for one geometry over another. However, we have also seen how the early advocates of relativity theory found themselves still fighting a rear-guard action against those who might be called "Euclidean *a priorists*," who still clung to an older style physics. Whitehead was absolutely *not* one of these, a fact he makes very clear on the very first page of text in R.

> It is this uniformity which is essential to my outlook, and not the Euclidean geometry which I adopt as lending itself to the simplest exposition of the facts of nature. I should be very willing to believe that each permanent space is either uniformly elliptic or uniformly hyperbolic, if any observations are more simply explained by such a hypothesis (R v).

Whitehead's use of Euclidean geometry was entirely a matter of *pragmatic* choice, a selection driven by what he had every reason to perceive as a simpler set of mathematical tools, rather than an *a priori* commitment regarding the structure of the world. On this point, it is important to remind ourselves that the tensor tools with which both Einstein and Whitehead formulated their theories were still extremely new in 1916 and 1922. As we saw in the previous chapter, both thinkers felt obliged to include large sections of introductory material on the subject in their works for a scientific audience that could not yet be expected to have any extensive familiarity with such things.[49] To contemporary eyes, educated in the (now) well developed tensorial techniques of GR, Whitehead's approach might scarcely seem simpler, with its confining geometrical structures, clumsy reliance on coordinate driven formulations, etc. But we must be careful how, and to what extent, we permit this contemporary perception to be read back into the history of the development of relativity. We need to ask ourselves why *Whitehead's* contemporaries gave so little attention to his theory.

To begin with, while experimental data played a part amongst a complex of factors leading to the widespread acceptance of Einstein's theories, it played no role whatsoever in the rejection of Whitehead's. There simply was no empirical basis upon which one theory could be preferred over the other. Quite early on, Eddington provided detailed confirmation of Whitehead's claim that his theory was mathematically indistinguishable from special relativity (Eddington 1924). Previously, G. Temple published an article working out more generalized results of Whitehead's theory for a space of uniform curvature (Temple 1923).[50] In this same time frame, we can find an unsigned review in *Nature* of R which is all the more glowing for the parallel comments of rather muted enthusiasm offered regarding a contemporaneous work on relativity by Arthur Eddington.

After this early spate of interest in Whitehead's work, there is hiatus of almost thirty years, carrying us well into our second period of relativity. But then, in the early nineteen-fifties, J. L. Synge generated a small resurgence of interest, when he expanded the range of comparison to GR

---

49. Section B of Einstein's 1916 paper on general relativity, pp. 120 – 142 in (Einstein *et al*). For Whitehead, this was part III of R, pp. 139 – 190.

50. I would also note in passing, that this article is often mis-cited in the literature. It is given citations by (Synge 1952) and the "Resource Guide for Physics and Whitehead" available on line from the Center for Process Studies at http://www.ctr4process/publications/pss/. Both of the citations are different, and both of them are wrong. The one given in (Will 1971b) is correct, though abbreviated. The one given in the bibliography here is correct.

by demonstrating that among a variety of then known effects – such as the perihelion advance of mercury's orbit, general rules for angular velocities of objects in a circular orbit, the deflections of light rays, and the measurable gravitational red-shift of light from the surface of the sun – the two theories either agreed outright, or differed to an extent that was not then measurable (Synge 1952, 303, 308). Two years later, C. B. Raynor added on Synge's results, which were only developed so far as to include static, spherically symmetrical systems, to non-static systems as well (Raynor 1954). An even broader comparison with GR was supplied by A. Schild in 1956, again showing the consistency of Whitehead's theory with that of GR to the limits of the available measurements, while also developing a method of generalizing Whitehead's theory into an entire class of theories of gravity, of which Whitehead's applied theory was a particular example (Schild 1956).

It is worthwhile to give some closer attention to Synge's work on Whitehead. For one thing, the first seven pages of Synge's above cited article are essentially given over to an apologia explaining why a mathematical physicist might yet find Whitehead's work an interesting and worthwhile puzzle, even though other types of physicists saw no good reason to give it a second thought. Synge states of Whitehead's theory that,

> It is not a field theory, in the sense commonly understood, but a theory involving action at a distance (propagated with the fundamental velocity c). To a 'naturalistic' physicist who feels that action at a distance is not 'the way in which nature works', the theory of Whitehead is repellent from the outset, but the 'mathematical' physicist is glad to add it to his collection of theories of gravitation, provided it is logically consistent, which it is (Synge 1952, 308).

The "naturalistic" physicists whom Synge mentions are those who are working from a more "conceptual" standpoint, motivated more by the intuitive, aesthetic, even philosophical aspects of their operative theory, while permitting the strict formal consistency of their ideas to slip and slide a bit, as needed. 'Mathematical' physicists are entirely consumed by the formal rigor of their theories, and do not trouble much over the intuitive appeal, or interpretive framework of said theory (Synge 1952, 304 – 307).

It is important to notice that the reasons Synge gives for the lack of consideration for Whitehead's theory are neither observational in character, nor are they related to the mathematical rigor of that theory. Rather, they are essentially aesthetic in nature, though Synge fails to note this explicitly. In any event, what catches his attention is that the "action at a distance" aspect[51] of Whitehead's theory is "repellent." However, Synge goes on to say that,

> If [the findings of the general theory of relativity are confirmed by observation], then the 'naturalistic' physicist is prepared to accept Einstein's theory as true (and this is in fact the present position). *But he might just as well accept Whitehead's theory, since its predictions in the case of* [rotation of perihelion, deflexion of light ray] *are the same as Einstein's (or practically so), while the case of* [the red shift in the sun's spectrum] *the same red-shift as Einstein's is given by an obvious interpretation of Whitehead's theory* (Synge 1952, 308, original emphasis. Phrases in square brackets are Synge's, albeit moved around slightly to make the quote more self-contained).

There is a kind of irony in Synge's claim, one that it seems unlikely that he was aware of. For it was only a year earlier, while also working with Whitehead's theory, that Synge made the following statement:

> . . . if the philosophy is only a wrapping for physical theory, then the mathematical physicist can take a savage joy in tearing off this wrapping and showing the hard kernel of physical theory concealed in it. Indeed there can be little doubt that the oblivion in which this work of Whitehead lies is due in no small measure to the effectiveness as insulation of what a physicist can in his ignorance describe only as the jargon of philosophy. The account of Whitehead's theory given in these lectures is emphatically one in which the philosophy is discarded and attention directed to the essential formulae (Synge 1951, 2).

Yet, with this sort of attitude in place, one cannot help but wonder: how, exactly, are the intuitions and models of 'naturalistic' physicists to be changed, so that ideas such as Whitehead's will no longer seem "repellent," when the very arguments needed to bring about such changes are viewed only as "jargon" and "insulation" to be stripped away? Implicit in the above – though, I would argue, emphatically real nonetheless – is the *failure* to recognize that the theory of nature which would characterize Synge's 'naturalistic' physicist is every bit as philosophical as that "wrapping" which he felt so thoroughly "concealed" the "kernel of

51. We will look a little more closely at this purported aspect of Whitehead's theory in later chapters. Suffice it to say here that, given the prominence within quantum theory of such action at a distance types of relations as non-local effects and quantum entanglement, to reject Whitehead's theory today on these grounds would be balking at a gnat while swallowing a camel.

physical theory" in Whitehead's work. Care needs to be exercised here, of course, because it is not immediately evident from the above whether Synge shares those views, or is merely reporting them as being representative of the opinions of many physicists. In any event, the real occlusion that is going on here is being performed by the *aesthetic* judgments – on the part of some physicists, at least – that set one class of philosophical notions apart, in a privileged position as "the way in which nature works," while viewing any concerted attempt to engage in a constructive re-evaluation of these notions suspiciously as "the jargon of philosophy."

Such, at least, was the situation up to the end of our second period of relativity. As a conclusion to this section, let me just say again that, throughout both of these periods, and well into the third, no empirical or mathematical evidence could be found that could disqualify Whitehead's applied theory of relativity as a viable alternative to Einstein's. Indeed, we can go farther and state, using the terminology from the previous chapter, that there were no effective *logical* grounds for rejecting Whitehead's theory either. Indeed, it was precisely such grounds that make his theory stand out, for at least Whitehead made a deliberate effort to provide them. It was not until 1971 that anyone could offer what at least appeared to be a specific, disconfirming instance, where Whitehead's theory appeared to disagree with observational evidence (Will 1971b). But even this cannot be taken as the final word on the matter. For even if it were to be shown that Whitehead's applied theory is finally, utterly, and absolutely disconfirmed in any present form, or any possible meaningful extension thereof, this cannot in any way be used to justify the failure to take his theory seriously during all the years prior to this hypothetical disconfirmation. Moreover, such a disconfirmation of Whitehead's *applied* theory would still leave open the issues he raises in the philosophical sections of his work, issues specifically relating to the possibility of a coherent theory of measurement in GR. Finally, we shall see in chapter five that, if such an absolute disconfirmation of Whitehead is indeed possible, it has yet to be produced: the argument given in (Will 1971b) is far from uncontroversial in its claims, and can be shown to have little or no bearing on Whitehead's general theory. Finally, and most tellingly of all, as we shall see below and in subsequent chapters, it doesn't really matter if Whitehead's particular theory has been disconfirmed, since there are numerous other perfectly viable theories that are fully consistent with Whitehead's uniformity criterion.

III. The Contemporary Scene in Cosmology:

Despite its widespread acceptance within the physics community, by the late 1950's to early 1960's general relativity had been in a kind of malaise for a good number of years. Most of this seemed to center around the lack of experimental evidence either for or against the theory. This was due to no failure of effort or imagination on the part of physicists, so much as to an absence of technology that could be employed in any then practical context. Clifford Will makes the following observation on this situation:

> By 1960, it could be argued that the validity of general relativity rested on the following empirical foundation: one test of moderate precision (the perihelion shift, approximately 1%), one test of low precision (the deflection of light, approximately 50%), one inconclusive test that was not a real test anyway (the gravitational red-shift), and cosmological observations that could not distinguish between general relativity and the steady-state theory. Furthermore, a variety of alternative theories laid claim to viability (Will 1993a, 8).

On this last point, by Will's estimate, something of a "cottage industry" (Will's phrase) of alternative theory creation had emerged such that by 1960 there were "at least 25 such alternative theories as found in the primary research literature between 1905 and 1960." (*ibid*). In the absence of a solid base of refined empirical evidence, these theories could scarcely hope to rise above the level of mere exercises, with the same forces working to drag the venerable GR down to the same level. The situation seemed so dismal that as late as 1962 Kip Thorne[52] was advised that "under no circumstances should he do general relativity, because it 'had so little connection with the rest of physics and astronomy'" (Will 1993b, 12). Only a robust and aggressive system of observational and experimental pursuits could hope to generate such a connection.

But, as a matter of fact, 1962 *was* late to be giving such advice. Already by 1960 a new wave of experimental and observational evidence

---

52. Kip S. Thorne is the Feynman Professor of Theoretical Physics at the California Institute of Technology, and is considered one of the leading experts in relativity theory. Among many technical works, he is co-author with Misner and Wheeler of *Gravitation*, a book which Carroll describes as having "educated at least two generations of researchers in gravitational physics," (Carroll, pg. 496) as well as authoring numerous popular books on the subject. Thorne was also Will's instructor at CalTech (Will 1993b, pg. 16).

was being developed which could finally provide the kind of vigorous experimental effort needed to keep a scientific discipline healthy. The years of 1959 – 1960 alone witnessed improved measures of unprecedented accuracy for things such as the gravitational red-shift and the deflection of star light around the sun, while entirely new tests involving such things as the precession of a gyroscope were developed (Will 1993a, 8). This resurgence of experimental effort led, in its turn, to a renewal of theoretical efforts. Most of this was, of course, directed toward refining the theoretical and mathematical understanding of GR, in an effort to derive new kinds of predictions which might yet be rigorously testable.[53] But this also led to a redoubling of efforts to formulate alternative theories of gravitation to that of GR. Depending on how one wishes to enumerate such alternatives – do theories that fall out as special cases of other formulations count as separate or not? etc. – Clifford Will enumerates over a half a dozen quite serious possibilities, all proposed and developed in the period from 1960 until 1979 (Will 1993a, 123 – 137). These theories, all of which were viable at the time that Will was writing, include several "bimetric" theories which employ a "prior geometry."[54] Which is to say, the geometry in these theories, unlike that in GR, is separated from the physical effects and relations of gravitation, so that geometrical structures are not dependent upon contingent physical relations. Such independence makes these theories of direct relevance to anyone concerned with Whitehead's criticisms of GR's lack of an independent geometry, and the consequences this has for a meaningful theory of measurement. We will have cause to return to the general idea of "prior geometry" and bimetric theories of gravity, and their relations to Whitehead's philosophical position in later chapters. The existence of such viable bimetric theories is the reason for our sanguine attitude towards Whitehead's own proposal.

One of the responses to this emergence of alternative theories was the recognition of the need for some effective means of comparing them, whether as one to the other or simply to GR itself. This led to the development of a kind of mathematical "machine" known as the "Parameterized Post-Newtonian" framework, or "PPN" for short.

---

53. One of the most famous and important examples of this is the so-called "Nordtvedt effect," whose details were discovered and worked out in the mid 1960's by Kenneth Nordtvedt while at Montana State University Bozeman. See, for instance, (Will 1993b, pp. 136 – 146).

54. These include theories proposed by N. Rosen from 1973 – 1975, P. Rastall's proposals in the years 1976 through 1979, and the work of Lightman and Lee in 1973. See (Will 1993a, pp 131 – 135) for discussion and citations.

Developed by Kenneth Nordtvedt and Clifford Will, the PPN formalism is the primary instrument for the comparison of theories of gravitation[55] (Will 1993a, 10). PPN might usefully be compared to the menu driven shopping systems one finds at many contemporary computer manufacturers. One goes to some particular manufacturer's website, and upon choosing to shop for a computer, one is presented with a list of choices that must be filled to create a customized computer order. Thus, one can select the type and speed of CPU, then go on to choose the type and quality of video card, then the CD or DVD, sound cards, etc. Each of the items on this selection list is a parameter which must be given some value or other (including possibly the value "none"). PPN is like this, only now one is shopping for a sequence of mathematical relations which can be assembled using the ten PPN parameters into a viable theory of gravity. And, just as one has a limited range of choices within each parameter in the selection of a computer (a CPU speed of 2.71828 GHz simply is not available), only a limited range and combination of PPN values will lead to a viable theory of gravity. We will be seeing more of the PPN formalism, and considering the kinds of choices necessary to, at times, shoe-horn theories into its format as we go on, especially in chapters three and five.

At this point, it is time to assess the ground we have covered, and use the foregoing material to explicitly open the question, How is one to select a theory of gravity and cosmology? What, ultimately, are the criteria to be employed in making such a selection? We need make no appeal to those approaches to the history and philosophy of science centered around rhetoric or politics, which might see such theory selection as emerging primarily from relations of individual or corporate power operating in an agora of ideas driven by ego and marketing skills,[56] to recognize that the process whereby a physical theory gains prominence within science has numerous and complex contributing factors. Even if we are satisfied that the only two factors that have a part in this matter are what might broadly be characterized as empirical adequacy and logical coherence, it is still the case that things are not simple.

Now, on the one hand, it scarcely seems to merit comment that, of course, issues of the empirical adequacy of a theory are of enormous

55. Specifically, so-called "metric" theories, which include Whitehead's. Observational evidence seems to have quite utterly excluded non-metric theories from any possible hope of verification. The nature of metric theories will be discussed in detail in chapter 3.

56. Nor is the above to be taken as suggesting that these factors play no role, or that the works which focus on these factors are of no value in understanding science.

importance in the conduct of scientific inquiry. Take as an example of this the earlier argument regarding the source and motivation of Einstein's theory of relativity. Emphasis was placed in that discussion on the role of symmetry and other aesthetic issues in the development of Einstein's theory. The particular experiment performed by Michelson and Morley in 1887 does not, on closer examination, seem to have proven such a decisive matter for Einstein's thinking as has commonly been supposed. But this can scarcely be construed as suggesting that empirical matters, either in general or particular, played no role at all in the development of relativity. As has already been pointed out, the opening paragraph of Einstein's 1905 paper is entirely devoted to the lack of theoretical *symmetry* in the treatment of readily observable, *empirical* facts. While Einstein never mentions the Michelson-Morley experiment in the 1905 paper, and on a number of occasions specifically states that it played no real part in motivating his theory, he does after all make a generic mention of the failed attempts to measure the Earth's movement in a stationary ether. So just because a famous experiment fails, on closer scrutiny, to live up the status assigned to it by cultural mythology, does not mean that empirical and experimental factors were absent altogether.

By the same token, issues relating to the logical coherence of a theory are constantly and explicitly being engaged at all levels of science. To be sure, if a theory is shown to suffer from various formal inadequacies, this hardly sounds the death knell on the theory. "Logical contradictions, except as temporary slips of the mind – plentiful, though temporary – are the most gratuitous of errors; and usually they are trivial" (PR, 6). But such errors are errors nonetheless, and certainly serve to indicate that further work needs to be done. Mathematical consistency is always a definite boon, and when it can be achieved in a theoretical framework it is always taken as a forcefully positive sign. Quite aside from the fact that such formal considerations show that the ideas within the theory link up in intelligible patterns which lend to the overall interpretation and clarity of the theory, consistent formalisms are efficient engines for driving the empirical testing of the theory. The more thoroughly mathematized a theory is, the more directly it lends itself to precise and rigorous experimental tests. By establishing a precise and consistently meaningful system of relations between ideas, the empirical consequences of those ideas can be distilled into an exceptionally refined set of testable possibilities.

On the other hand, however, we already see hints in the above of how conditions such as those of empirical adequacy and logical coherence are not "closed" or "self-contained" in the sense that their meaning, relations, and consequences can be compactly pigeon-holed into neatly bundled packages: for instance, *mathematical* developments of a theory lead to unanticipated *empirical* claims. But the problems cut far deeper than just this superficial lack of closure in these criteria. For one thing, the very notion of empirical adequacy could scarcely be any more vague. Of course a scientific theory must be solidly anchored in the empirical facts; but just which are *the* facts to be taken as relevant, what exactly do they mean, and how do we draw them out from the background of equally real events which nevertheless have somehow not merited qualification as anything other than "noise"? We have already seen how Michelson, for decades after his famous experiment with Morley and almost to the end of his life, regarded his attempt to measure the Earth's movement through the ether with his interferometer not as a crucial experiment in cosmology, but as an out and out *failure*. In this context, it is also appropriate to mention Eddington's famous expedition to measure the displacement of stars appearing near the limb of the sun (presumably due to gravitational effects predicted by GR) during a solar eclipse. In order to declare the expedition a successful test of Einstein's theory, Eddington had to ignore the majority of the photographic images taken during the eclipse, choosing to count as "facts" only those plates which gave – and here, one must also add, gave only *on average* – results that could be interpreted as consonant with those predicted by GR.[57] As Collins and Pinch point out, Eddington was not working according to a "double blind" set of rules such as we might find in a medical test; he was strongly motivated by his prior theoretical commitments to find the evidence he was seeking (Collins and Pinch, 49 – 50).

Similar examples can be multiplied without effort. The point that must be acknowledged here is that events do not wear their own interpretations "on their sleeves," as it were. "Facts" are only recognized as such when events have been successfully and appropriately embedded within a system of interpretations which can selectively acknowledge some of those events as facts, and then say just what kinds of facts they are, by relating them together as a meaningful whole.[58] And empirical adequacy

---

57. On this, see (Collins and Pinch 1993, especially pp. 48 – 52). More generally, this work examines a variety of scientific experiments where the empirical evidence proved to be anything but transparent in terms of its appropriate interpretation, including (in relation to our concern here with cosmology) further material on the Michelson-Morley experiment, and the attempts to detect "gravitational waves."

58. This is so even if one believes in an ontologically and objectively real "Truth". For

cannot be the only criterion used in the selection of its own interpretive scheme. As we have seen, philosophical and aesthetic factors come into play as well – and as a result, some theories get ignored as "repellent" despite the fact that they are lacking in neither their empirical adequacy nor their logical coherence. Such additional factors are always and inevitably active players in the determination of when we feel vindicated in saying, "Aha! Now we know," rather than just, "Oh well, that did not work."

And in a similar vein, logical coherence invokes bigger issues than just the relatively provincial matter of a theory's mathematical consistency. Ideas link together in a variety of ways, and not all of those linkages can be reduced to a mere formal absence of contradiction. Indeed, the self-same aesthetic considerations we've encountered above not only come into play here as well, it is in the arena of ideas that such play is most prominently carried on. We have already seen that aesthetic issues exercised prominent roles in the development of relativity and the rejection of Whitehead's proposed approach. Quite aside from a theory's logical consistency, there are issues of its degrees – and let us be emphatic here, *kinds* – of symmetry. For while symmetry itself can be given a mathematically consistent development in the form of the mathematical theory of groups, this by itself can hardly tell us where, which, and how those group actions and structures – those symmetries – are to be found, represented, and interpreted. The *selection* of which symmetries are important does not force itself upon us; it too, is a manifestation of a variety of factors, including aesthetic and philosophical ones.

Indeed, as we shall see (especially in chapter six), the issue of which symmetries are to be valorized as "true" ultimately stands at the crux of the matter between Whitehead's and Einstein's respective formulations of relativity. For Whitehead, the inhomogenous metrical relations of GR, which are read off in that theory as the curvature of space and time in response to the contingent distributions of matter and energy throughout the universe, and which he criticizes as a failure of uniformity in GR's geometry, are in fact a break down of *symmetry* in GR.

It is ironic, at the very least, that the language of symmetry and the mathematics of GR, in the form of the theory of Lie groups[59] and Riemannian spaces, did not come to be formally wedded together until some four years after the publication of R, in the seminal work of the great

only as the (presumably, uniquely) correct interpretive scheme is applied to events can "The Truth" be recognized as such.

59. Named for the Norwegian mathematician Sophus Lie (pronounced "Lee"), 1842 – 1899.

French mathematician Elie Cartan.[60] By this time, Whitehead had ceased to invest any serious effort in the mathematical development of his applied theory of relativity. At the same time, the physics community was no longer showing even the minimal levels of interest in Whitehead's theory as was to be found in Temple's brief paper of 1923, or Eddington's 1924 note in *Nature*. By the time Synge returns to Whitehead's theory in the early 1950's, the only part of that theory that is given any treatment is the mathematical part; the philosophical motivations are ignored.[61] It is, of course, these philosophical motivations which clearly inform us of the general structure of what Whitehead would consider to be an adequate theory of relativity. It is here that one recognizes that Whitehead's uniformity criterion is not exclusively invested in the special form of the applied theory he offers up in R as an *exempli gratia*. On the contrary, with the advantage of what is now – *for us* – hindsight, we come to see that what Whitehead's general theory requires, and his special theory exemplifies, is a high enough level of *symmetry* in the metrical structures of space so that measurement relations will have a sufficiently robust form as to be meaningful. As we have observed above, these symmetries will express themselves mathematically in the group theoretical structures which manifest themselves in the metrical forms on the space. Consequently, Whitehead's uniformity criterion, far from being a mere piece of "philosophical jargon," is a formally definite feature that, by Whitehead's argument, should be present in any adequate physical theory. And, in this instance, "adequate" means a theory with a coherent explanation of measurement.

As nearly as this author is able to determine, this relation between the mathematical *symmetry* of the space of theoretical physics and Whitehead's "uniformity" criterion has never been noticed in the literature. Consequently, it most emphatically is a theme to which we must give our

60. This in two papers from 1926; see the bibliography in (Helgason 1964) for the references. The original papers can be found in the *Oeuvres Completes* of Cartan's work. Helgason's work, which is the standard reference on this material, is quite appropriately titled *Differential Geometry and Symmetric Spaces*.

61. It should be mentioned that such a response was hardly new with Synge. Whitehead was present at the reading of Temple's paper at the Physical Society of London, and his comments (as well as those of others) are included in the "minutes" at the end of the published paper. It is clear from reading these that, while he found Temple's treatment of the mathematics admirable in its competence, Whitehead was even then finding it necessary to re-emphasize that and how his theory differed in a fundamental way from Einstein's, not just mathematically, but philosophically as well. See the endnotes to (Temple 1923).

particular attention in the coming chapters.

CHAPTER THREE
Metrical Theories vs. Theories of Measurement

> *...the complete explanation of number awaits an understanding of the relevance of the notion of the varieties of multiplicity to the infinitude of things. Even in arithmetic you cannot get rid of a subconscious reference to the unbounded universe. You are abstracting details from a totality.*
>
> Alfred North Whitehead, "Mathematics and the Good."

I. Structures of Metrical Theories:

Having very lightly glossed the conceptual framework of relativity at the opening of the previous chapter, it is time now to look a bit more carefully at the mathematical structures which are the formal expression of those concepts. Our intention here is to take a look "under the hood" of GR, and learn a little more about some of the major components of that theory. We want to get a sense of the "geometrized" – which is to say, standard – interpretation of that theory and gain more of a sense of how those pieces work together. From that vantage point, we will then be able to get a more rigorous sense of how it is that a successful metrical theory such as GR can nevertheless fail to supply us with a coherent theory of measurement. As we shall see, the former can provide us with a "machine" which, given the appropriate "inputs" will provide us with numerical values that can be translated into empirical tests. On the other hand, it takes for granted precisely those *logical* relations – such as congruence and or equality – which a theory of measurement seeks to explicate. This latter type of theory, in its turn, has little to offer (directly, at least) when it comes to generating numerical values that are possessed of any observational content. Knowing *why* our looking at things *can* be meaningful does not provide us any substantive clues as to *what* those (numerical) meanings might actually be.

As before, our approach will be highly intuitive and informal by mathematical standards. However, we will be rolling our sleeves up a bit this time, and actually looking at some of the mathematics. Our aim here will be to gain a working familiarity with both the metric tensor and the Riemann curvature tensor. We have actually seen a "picture" of this latter in chapter one, but we went no further in our understanding of it than to get a very slight sense of how the switching of a couple of the indices could formalize a relation that could in turn be interpreted as curvature. This time around, we will be digging down into the individual symbols of both the Riemann and metric tensors. This means getting an outline understanding of the kind of mathematical structure in which those tensors operate. This will at last provide us with a more realistic sense of just how geometrical GR really is, and why such an interpretation is so intuitively plausible and attractive.

When we move on to section II, these additional technical details we have developed here will serve us as we look more carefully at the functioning of the PPN formalism. This will allow us to understand more fully how competing (or, at least, potentially competing) theories of gravity are actually compared. This, for its part, will also help make explicit the kinds of structures, relations, and interpretations that are presupposed in order to make such comparisons “work.” Indeed, throughout both these sections and the third, one should be constantly asking oneself, “what sort of underlying reality is being assumed here, that these mathematical structures are the most appropriate representatives?”[62] Finally, in section III, we will turn to an examination of those logical relations needed for any coherent theory of measurement, and assess in more detail how the theoretical commitments of GR actually block the fulfillment of those requirements.

To begin then, we have already mentioned tensors quite a bit, and even looked rather cautiously at a few aspects of how they operate. It is now time to dig more deeply into this idea, so that we might gain a more concrete sense of what a tensor is and why it seems so natural to think of it as a geometrical entity. For this, however, we must back up a bit, and look at some of the component ideas that went into the development and origination of the tensor. Now a tensor, as we have already noted, is an entity with various associated indices, some of which can appear as subscripts, others as superscripts, and sometimes some of each, but which are together the mechanics underlying the “book-keeping” function of the tensor. The tensor itself, stating the matter baldly, can be defined as “a multilinear map from a collection of dual vectors and vectors to ” (Carroll, 21). Now, at first blush, this definition might appear practically uninterpretable to anyone not already indoctrinated in the arcana of differential geometry. Moreover, even amongst the initiated, “for tensors arising in applications or from mathematical structures it is rarely the case that the multilinear function interpretation of a tensor is the most meaningful in a physical or geometric sense” (Bishop and Goldberg 1980,

62. This question gets addressed explicitly in chapter four.

79). Nevertheless, precisely because we want to avoid prejudicing the matter by naively accepting the "geometric metaphors" which ordinarily govern the development of these ideas, it will serve our purposes to take the more "abstract" point of view on what a tensor is. And it is still possible to tease out information from the above definition that can direct our attention to some of the essential characteristics of the tensor. Two facts in particular present themselves in the above: first, a tensor is a kind of thing (a "multilinear" map) that, second, relates two sorts of other things together – vectors and dual vectors on the one hand, and the real numbers on the other.

So, if we begin by simply noting the adjective qualifying the nature of the mapping function of the tensor, we might suppose – correctly, in this case – that a "*multi*linear" map is in some sense or other a multiple iteration of an ordinary *linear* mapping. In its turn, linearity is not a difficult idea to grasp. For a function or mapping to be linear means both that the objects the function maps and the function itself can be combined with, respectively, other objects from the collection of things mapped or other appropriate functions[63] in a straight-forwardly number-like manner. Thus, if "f"and "g" are such appropriate functions, "v" and "w" are among those objects mapped by f and g, and "α" and "β" are ordinary numbers, then some of the basic rules of linearity are the following:[64]

(v and w behave in a number-like fashion)

(1) $f(v + w) = f(v) + f(w)$ (the map of the sum is the sum of the map.)

(2) $f(\alpha v) = \alpha f(v)$ (the map of the multiple is the multiple of the map.)

(f and g behave in a number-like fashion)

(3) $[f + g](v) = f(v) + g(v)$

(4) $[\alpha f](v) = \alpha[f(v)]$

The two sets of rules combine in the obvious ways, again maintaining the number-like relations between the objects that are being mapped (the "v"

63. Which, in this case, means that the mappings are from and to the same collections of things. In our case here, that means from vectors and dual vectors on the one hand to the real numbers on the other.

64. See, for instance, (Bishop and Goldberg, chapters 1 and 2), for a more extensive discussion of these topics.

and "w") and the objects doing the mapping (the "f" and the "g").

Two facts worth noting here are that, in the first place, these different "things" (objects mapped, and objects mapping) are only obeying number-*like* rules. They are not, or at least need not be, numbers themselves. And in the particular instance here, where the subject is the nature of tensors, the objects we are dealing with are vectors and their *duals*. As previously mentioned in chapter one, the most basic notion of a vector is as a geometric entity which has both a direction and a magnitude. We encountered some of the number-like behavior of vectors there when we imagined going from the corner of Mill and Illinois to Oakland and Main. Each leg of our journey had a direction (for instance, North from Mill) and a magnitude (the distance to Main St.). We could take the sum of two vectors – say, North to Main plus West to Oakland – and have a meaningful result; namely, the corner of Oakland and Main.[65] Moreover, had we a different destination in mind, we could have taken one of our "vectors," North to Main for instance, and simply doubled it, which is also known as multiplying the vector by a "scalar," placing us around Illinois and Willow. So both the addition of vectors and the multiplication of a vector by a number is sufficiently number-like to qualify as linear.

Now, as just mentioned, vectors have a "magnitude," a numerical value that tells us that vector's "length." This length is typically determined by means of a function which (in Euclidean spaces, at least) is known as an "inner product." An inner product, as it happens, is a rather general function that can be applied to *any* two vectors. The numerical value the inner product provides, when applied to any two arbitrary vectors, is in essence a numerical value for how much "in common" those two vectors have. In the Euclidean context, this inner product is often represented with a "·," (sometimes also called a "dot" product, for the obvious reasons) so that the inner product of two vectors such as "v" and "w" would be

$$v \cdot w = \alpha$$

where "α" is the numerical value telling us how much v and w are "alike." When, however, this inner product is applied not between two different vectors, but the same vector, as in

$$v \cdot v = \alpha$$

since the v is exactly "like itself," the resultant α is the *square*[66] of the

---

65. This sum of vectors is a different matter from the parallel transport issue – our "compass vector" – discussed in that chapter.

66. Much as a number multiplied by itself equals its own square. Since, however, a vector is not a number, it takes something like the inner product to produce a numerical value associated with it. The square-root of this value can then be taken

length of v.

General Relativity tells us that every point of space is equipped with something like this inner product. This is the "fundamental," or "metrical" tensor. Because this tensor always works with two (not necessarily different) vectors, giving a metrical relation between them, it always has two indices. Tradition has come down to us that this metrical tensor is represented as "$\mathsf{g}_{\mu\nu}$." To ensure clarity here, we will always represent the metrical tensor as a *sans serif* "g."

The second of the two previously mentioned points to notice, is that the functions which are mapping the vectors are themselves rather "vector-like" in their behavior. Just like two vectors can be added together to form a composite vector, so our functions f and g can be summed to form a composite function. And, just as a vector can be multiplied by a scalar (i.e., a number), so can our linear functions. Indeed, since these functions are mapping their relevant objects into the real numbers, multiplying the function by a number is the same as multiplying the resultant real number produced by the function operating on one of its objects by that self-same number. Such an operation always makes sense by definition, whatever else the functions or the objects functionally mapped might be. The very nature of these relational structures invites one to move from a merely analogical view of these functions as "vector-like," and look upon them instead simply as a kind of vector. The entire collection of such functions comprises a "space." When the objects operated upon by these functions are themselves vectors – "real" vectors, one is tempted to say – then the space of functions stands in a special relation to the space of vectors. The function space is "dual" to that of the vector space, and the functions themselves can be viewed as "dual vectors." This duality is so profound that one can entirely invert how the relations are viewed. Rather than seeing the dual vectors as functions which map the "actual" vectors into the real numbers, the "actual" vectors can be treated as the functions which operate to send the dual vectors (what we had been viewing as the "actual" functions) again to the real numbers.

Already this gives us enough to return to our initial definition. If a tensor is a "multilinear map" then it is itself a kind of linear function, obeying the number-like rules of linearity, only doing so in a manner that is multiply indexed across a list of objects all at once. At the same time, the objects it is mapping are collections of vectors and dual vectors – which is to say, functions on vectors – into the real numbers. So it is acting linearly on linear objects with a resultant value from a structure (the real numbers)

---

to derive the true value of the magnitude or length.

which is itself fundamentally linear. And certainly it is the case that, if one focuses exclusively on the purely structural relations involved, then the temptation to interpret all of these mathematical objects geometrically can scarcely be resisted – why would one even want to? The answer to this last, obviously rhetorical question, is that we are trying to make the "space"[67] here to honestly evaluate a scheme of nature that stands in genuine competition to that presupposed by general relativity. In this particular instance, it is especially useful to note that the interpretation of functions on vectors as "dual vectors" geometrizes objects which, unlike vectors proper, have no immediate spatial characteristics. Such a move is predicated exclusively upon the formal analogies of certain structural features of the relations amongst these functions to formal, structural features of more evidently geometrical objects. The converse move of seeing vectors functionally, as operators mapping "dual vectors" into the real numbers, simply reinforces the pan-geometric interpretation of objects which were, initially at least, quite different.

When one is dealing with pure mathematics this kind of reductive move hardly seems objectionable. Mathematics is all about formal analogies between purely structural relations. So the building of such analogies, as well as the cultivation of the intuitions which can sense where and how best to do so, is an eminently admirable and appropriate course of action. And the fact that these vectors and functional relations are being mapped into the real numbers already brings to the fore the theoretical possibility of measurement. *If* one can associate these mathematical objects with their appropriate physical analogs, and if, in turn, one can gather these vectorial and functional analogs together under the aegis of an acceptable tensorial mapping, then the resulting real number will effectively be a measurement. If that tensor – which, as a mathematical object quite beyond the specific physical realities associated with any particular vectorial or functional relations, already enjoys a high level of generality – can be brought into play with other such vectorial and / or functional relations, then one now has a second number which can be compared with the first, and the possibility of measurement has become a reality. It is precisely for this reason the the "fundamental" tensor, which in the scheme of GR is to be found at every point of spacetime, is also known as the *metrical* tensor. This is the number producing multilinear engine which drives GR.

But make no mistake here: this *is* a kind of reduction. Functional relations of vectors are not the same things as vectors, however many

---

67. Mea culpa – but this sad little pun also serves to indicate the universality of the temptation to engage in spatial metaphors.

interesting formal morphological features they might share with the objects they are mapping. Aspects of these functions are being ignored so that they may be converted from analytical, relational schemas into geometrical "things." And while such a move has a number of advantages within the context of pure mathematics, once we have moved our attention to the realm of physical reality and nature, this methodology can no longer be allowed to pass without question. To do so would be to simply collapse our philosophy of nature into our mathematics, with little or no regard for either the full range of our experience or the fundamental requirements of our knowledge, in coming to understand the very nature we presume to mathematize. But it is also the case that raising these questions cannot of itself be taken as answering the questions raised. A cautious skepticism toward a carelessly easy geometric interpretation of the functional and tensorial forms found throughout any even remotely viable theory of gravity and spacetime must not be construed as suggesting these theories have no appropriately geometrical content. All that we might hope to claim at this point is that the ease with which our analogical intelligence allows us to see such geometrical possibilities must not be allowed to run away with the issue.

These periodic cautionary reminders are a nuisance. But it can hardly be overemphasized just how extensively geometric GR is by any account. Tensors, as we have seen, are mapping combinations of vectors and functions on vectors into the real numbers, giving us the numbers which are the content of a physical measurement. But, as soon as we ask ourselves, "Where are these collections of vectors, etc., being mapped *from*?" the robustly geometric character of our enterprise once again forcefully asserts itself. This brings us to a mathematical structure that, while not as old an idea as that of a vector, in the context of contemporary cosmology is easily as fundamental, namely that of a "manifold." A manifold is a generalization of the traditional idea of a Euclidean type of geometric space. In the Euclidean geometry of our high school days, the entire space could be analytically represented by a single, univocal system of Cartesian coordinates. In a manifold, however, this need no longer be true. All that can be categorically asserted about a manifold is that every point admits of a *local* coordinate system, but there is nothing to guarantee that this local coordinate system can be extended in any sort of global manner; indeed, such an extension is, in general, quite impossible. In other words, at any point in a manifold, if one narrows one's vision sufficiently, the region around that point will appear to be a Euclidean space of some given dimension. But this narrowly defined image that can be found around any given point by itself tells us nothing of the overall structure of the manifold itself.

Once again, there are a couple of points that deserve to be noticed here. To begin with, the above description should be provoking a kind of *deja vu* of the previous chapter's discussion of the transition from a Newtonian global frame of reference in the form of absolute space and time, to local Lorentz frames found within relativistic cosmology. This, of course, is no accident. The idea of a manifold, originally developed by Riemann and others in the latter half of the nineteenth century, was chosen as precisely the tool needed to express this separation between the local and global structures of space, and thereby set relativity on a solid mathematical footing.

The second thing to notice here is the oddly pragmatic nature of the above. A manifold is "locally" Euclidean; every point admits of a coordinate system in a "small enough" neighborhood around that point. But there is nothing here to tell us what constitutes "local" or "small enough." And this is the case even within the purely mathematical development of manifolds. All we know about a manifold is that, for each point, there is – by definition – a "small enough," localized area "near" that point that can be made arbitrarily similar to a Euclidean space. In pure mathematics, this quality of "small enough" is generally a matter of stipulation. One just declares that, for instance, "U" is a neighborhood of the point "p," and it goes without further saying that it is as "local" as the circumstances require. Those circumstances can be quite extensive. For instance, Euclidean space is itself a manifold since, indeed, there is a neighborhood around every point of the space that is Euclidean. It so happens that that "local" neighborhood can be as big as the entire space itself, but it is still a "neighborhood." As has already been mentioned, in most mathematical manifolds, the scope of "locality" is not so all enveloping. Yet it will be there, in order for the space to be considered a manifold at all.

But in physics one has the initial sense that we cannot – or at least should not – casually lean on such *ipse dixitism* to resolve our theoretical complexities. And indeed, we do not *casually* lean on such techniques. But lean on them we nevertheless do, even if only to a degree. Absent the assumption of these minimal, local uniformities, a theory of physical space would be so utterly devoid of any relational coherence that one can scarcely imagine what possible empirical content such a theory might have to offer. It *must* be possible to compare one point of space with another for any theory of space – especially physical, empirical space – to be possessed of any scientific meaning at all. And to declare that a point of space was devoid of any degree of Euclidean structure in any region about that point, no matter how limited or small that region might be, would effectively extract said point from any possible basis of comparison to other points (such as those within our immediate, experiential space) which were otherwise so endowed. For the very possibility of comparison would, by itself, suffice to equip that first point with at least some aspects of the more Euclidean points to which it was being compared. Else, what about the point could even be compared? So while the assumption that space embodies the fundamental structure of a manifold commits us to the "merely" pragmatic assumption of a minimal level of uniformity, it is hardly conceivable what the empirical gain would be for assuming otherwise.[68] We will revisit this theme in the final chapter.

In the meantime, continuing the development of the pragmatic stipulations made by general relativity, the next layer of these is to be found in the presumption of the kinds of connections that exist between and amongst points of space so that effective, metrical comparisons can be made between them. The assumption that every point of space is endowed with a tensorial function (the metrical, i.e. fundamental, tensor) means that physical space is appropriately akin[69] to a very special kind of manifold, a pseudo-Riemannian manifold endowed with a Lorentzian metric. (See Bishop and Goldberg, 120 ff. for more on these terms, and what they mean.) This is a richly structured type of space, one which permits of the relatively smooth comparison of the metrical qualities of one point in space with another, by means of a kind of differentiation that is the purpose built extension of the older notion of a partial differential equation.

Recall our example from the first chapter of our very rough and ready story of different paths taken to the same corner of town, revealing an unexpected vectorial difference in the reading from our compass. As stated at that time, the example given was a massive simplification of the ideas employed in GR to develop the concepts and consequences of curved

---

68. It must be added that the "localness" involved can be extremely local indeed. Thus, for instance, GR not only makes possible, but even necessary the existence of "singularities" ("black holes") which – almost – stand out of all relation to the rest of space. Yet, even here, it is the uniformity of space which makes it possible to detect these singularities, since it is this very uniformity which creates the meaningful background against which the dramatic effects of black holes theoretically make themselves known.

69. The cautiousness of this expression is the author's, and does not seem to be typically adopted in the physics literature, where it would more commonly be said that space simply *is* a Lorentzian type of pseudo-Riemannian manifold.

spacetime. The idea was that if an appropriate vector could be "parallel transported" from its starting place to a more distant point along alternative paths using a sequence of small steps (the "little bits" of differentiation), then one could use the different results on the test vector from the alternative paths to make a determination of the curvature of the intervening space. Now, as it happens, the standard partial differential equation is inadequate to this task on two accounts. In the first place, partial differentials always commute, so the order in which they are taken is incapable of revealing the kind of information we are seeking, information which differs precisely because of a difference of order. Secondly, and at least as importantly, partial differential equations are meaningful only in the context of a specific coordinate system. In other words, P.D.E.'s by themselves do not behave in a tensorial manner, as invariants between different coordinate frames. Since coordinate frames are only valid within the local context of any particular point, they cannot be used in determining global structures.

However, it should come as no surprise that there is a kind of differentiation available to us that corrects both of these faults. By taking a select system of P.D.E.'s and adding to them an appropriate set of "corrective factors," what we then have is known as a "covariant differential." The covariant differential is often symbolized as "$\nabla_\mu V^\lambda$" where the specific "λ" component of the vector V is being differentiated in respect to *every* "μ" component of the vector (we are comparing how the one component of the vector differs from all the others). Like the P.D.E.'s it is built from, the covariant differential retains the structure of a "multi-dimensional" form of differentiation (see the discussion in chapter one), but with the corrective factor added, it is now simultaneously sensitive to the order of differentiation and yet indifferent to the coordinate scheme employed.

The development of these ideas originally stemmed from the study of manifolds with a densely metrical structure, such as was ultimately used to characterize space in GR, with every point of the space equipped with a metrical tensor function. With these tools in place – a robust metrical structure and a covariant derivative – mathematicians were able to abstractively derive the notion of a "connection" on a particular space. The idea of the connection came straight out of the "corrective factor" which had been added on to make the covariant differential work. These correction factors were essentially systems of nXn matrices,[70] where n was

70. A matrix is a table of rows and columns, in which each entry is filled by a number, and the space in the entry can be identified by its respective place in the system of

the dimension of the entire space, with one such matrix for each component μ in the particular vector. Taken in their own right as the "connection" of the space, they provided the systematic rules whereby each point's relations to others in the space could be set out, and matters of curvature (both locally around a particular point, and / or globally throughout the entire space) could be schematically represented.[71]

Once distilled from the historically original ideas of metrical relations and covariant derivatives, these relations came to be inverted, with the connection introduced axiomatically and studied as a structure in its own right, and the covariant derivative built up from there. The connection itself is commonly represented as a tensor-*like* form (but, it must be noted, by itself it most definitely is *not* a genuinely tensorial invariant) called the "connection coefficient."[72] This is typically symbolized in contemporary mathematical notation with a capitol gamma, equipped with one upper and two lower indices, such as with "$\Gamma^{\lambda}{}_{\mu\sigma}$." A quick glance back at the Riemann curvature tensor (which appeared in chapter one) should indicate how central the connection coefficient's function is in the representation of curvature on a metrical manifold. Furthermore, the connection symbol appears in the definition of a "straight line" on the manifold, known as a "geodesic." That the connection coefficient is central to the definition of a straight line should not come as a surprise, since the relations of geodesic – i.e., straight – paths and spatial curvature are directly complementary to one another.

Using the connection coefficient, the covariant derivative can now be conveniently defined as:

$$\nabla_{\mu}V^{\lambda} \;=\; \partial\mu V^{\lambda} + \Gamma^{\lambda}{}_{\mu\sigma}V^{\sigma}\,.$$

The first term on the right of the equals sign is just the partial derivative of the specific λ component of V in comparison to all of its μ components. Recalling that the "corrective factor," i.e., the connection coefficient, was just a system of nXn matrices, what the gamma is adding is a series of matrix multiplications of each component of V (now indexed with the σ in order to show that this multiplication is independent of the specifics of the

rows and columns. An nXn ("n by n") matrix is one with an equal number of rows and columns, also called a square matrix (for the obvious reasons). Matrices are really the mother of all studies of linearity: they can be added, multiplied, etc. Pretty much any basic text on linear algebra will begin with the study of matrices.

71. For more technical details on this subject, see Bishop and Goldberg, pp. 219 ff.

72. Also as the "connection symbol," or as the "Christofel symbol."

earlier P.D.E. term) *connected* to the partial derivative by the "μ"and the "λ." I.e., it is not just any set of nXn matrices, but the μ set along the λ component multiplied with all of V's components, using the "σ" index (Carroll, 94 – 5).

These indices tell us nothing about the nature or structure of these matrices. Recalling our earlier discussion about the metrical tensor, it was said that the metrical tensor basically took the inner product of two (not necessarily distinct) vectors to give a numerical measure of how much "in common" those vectors had, or it could operate on a single vector (twice) and give that vector's magnitude. If we consider this latter approach, we can limit the operation of the metric tensor to individual components of a vector – which means treating those components as vectors in their own right – and get a "sub-measure" of the vector at any given component. Thus, for instance, $g_{\mu\nu}V^{\nu}$ will serve to give a magnitude of the ν component of V. Furthermore, it turns out that the behavior of g is very much like that of a matrix in its own right, which means that it has an "inverse matrix"73 represented as "$g^{\nu\mu}$." With this in mind (and leaving the vector component V^{ν} implicit in order to keep things simpler), we can now make the connection coefficient a bit more explicit. The connection coefficient is defined using g as:

$$\Gamma^{\sigma}{}_{\mu\nu} = \tfrac{1}{2}g^{\sigma\rho}(\partial_{\mu}g_{\nu\rho} + \partial_{\nu}g_{\rho\mu} - \partial_{\rho}g_{\mu\nu})$$ (Carroll, 95 – 9).

Remember that the indices are arbitrary book-keeping devices: they can be exchanged and or substituted at will, as long as one is consistent and the relations remain unchanged. Secondly, note that only the "v" corresponds to the component of the vector that the connection coefficient is operating on (and which, again, we've left implicit here). All the other indices, while they must track together in terms of, for instance, the "μ" column of one g versus the same row of another, are either referring back to another P.D.E (the $\partial_{\mu}V^{\sigma}$ from the covariant derivative) or are entirely for "internal" book-keeping purposes of the equation (for instance, the "ρ" index).

While all of this might seem a bit dizzying, the important thing to realize is that it turns out that the systems of nXn matrices are composed from the metrical tensor! The reader is invited to plug the above definition back into the Riemann tensor from chapter one, taking care to keep all of the indices straight, adding and modifying these as needed for the sake of consistency, and then contemplate the appropriateness of calling the

---

73. I.e., one which, when multiplied with the original matrix gives an "identity" matrix, one with 1's down the diagonal and 0's everywhere else.

metrical tensor "fundamental." Even if one is content to expand the Riemann tensor in one's imagination alone, it should also begin to be apparent how, if the metrical tensor depends exclusively on the distribution of matter and energy in the universe, the curvature of space is itself absolutely infected by the contingencies of those distributions.

Retreating to a somewhat more conceptual (as opposed to formal) level, the covariant derivative itself can be thought of as measuring the "rate of change" of various vectorial and tensorial quantities as one moves along some path in space from one point to another. Because of the connection coefficients, this notion of "rate of change" is meaningful, in that each point's vectorial and tensorial quantities can be compared to those of neighboring points and these quantities can be taken as standing in an appropriately similar relation at these neighboring points, such that while some of the quantitative values might change, the underlying meaning of the vectors and tensors being compared is essentially the same. Indeed, one can treat of these vectorial quantities at different points as though they were the representations of a single vector which has been "parallel transported" from the first point to the second. For a vector to be parallel transported means that for each "infinitesimal"[74] step along some path from the first point to the second the "same" vector can be compared using covariant derivatives to ensure that each instantiation of the vector is "parallel" (pointing in the "same" direction, given the appropriate sense of "sameness") as those to either side of it along the specified path.

However, within a curved spatial manifold, such as physical spacetime is conceived to be, different paths from the same starting point to the same ending point which – individually, at least – parallel transport the "same" vector, will in general result in this vector pointing in very different directions. It is such differences which are the very manifestation of the curvature of space. Consider, for instance, the surface of a globe. Now, if one can step away from this surface, moving out into the three dimensional space in which the sphere is embedded, one can directly observe that the sphere's surface is curved. But suppose one does not have this option, suppose one was ineluctably stuck on this surface. How would one determine if the surface one was on was truly curved or flat? Well, using a globe of the Earth as a working tool, we can take a toothpick and

---

74. I.e., a step too small to make any measurable difference within one's operational context. That there might be such things as infinitesimally small distances in reality is generally not discussed in the physical literature. As we shall see, the RNI assumption effectively commits one to such a claim, whereas Whitehead's philosophy of nature treats such things as extreme abstractions.

place one tip of it so that it just touches the globe at some point along the equator, while the length of the toothpick, though it is not touching the surface, nevertheless lines up directly with the line of longitude through the point at the toothpick's base. The toothpick is said to be "tangent" to the surface of the globe at the chosen point of the equator. We can then parallel transport the toothpick along the line of longitude until we reach the north pole of the globe. The base of out toothpick will now be tangent at the pole, and pointing in some specific direction (the body of the toothpick will still not be touching the globe). Noting this direction, we can return our toothpick to its starting position along the equator, pointing again in its original direction.

Now, however, let us first move the toothpick along the equator some substantial distance, say a quarter of the way around the globe. If we take care to keep the toothpick lined up with whichever line of longitude runs through the new tangent point at the base of the toothpick, we can again move the toothpick along this line of longitude to the pole. We will again be certain that we are parallel transporting the toothpick, because the lines of longitude on the globe *are* parallel lines on a *spherical* surface. But now, our toothpick is pointing in a direction that is 90° from the ending position of the original move. So even though the same toothpick starts from and ends at the same points on the globe, because it follows a different path to reach its destination even though it remains parallel to itself on each of the different paths, its finishing direction is completely different (See Carroll, 102 ff).

This process is the same for a space of any dimensions, not just a two-dimensional surface such as that of a sphere, provided there is *some* kind of rule of parallels that makes sense for that space, and a technique for characterizing a tangent vector within that space. This means that there must be some kind of rule of differentiation that makes it possible to say that, between any two neighboring points (where, again, "neighboring" means "close enough to make things work") it is possible to say that two tangent vectors point in the "same" direction, where this concept of "same" is at once functionally useful without disrupting other functionally viable structures in the space. This balance of functional characteristics is of central importance in this determination of what does or does not qualify as "parallel." One might otherwise as easily declare the lines of latitude on the surface of a sphere to be the parallels, rather than the longitudinal lines, arguing that in this way the "parallel" lines would always be a fixed distance away from one another, rather than intersecting as lines of longitude do. But such an arrangement would cause too many other relations to break down, so that despite its initial plausibility, the "latitude as parallel" notion does not lead to a satisfying system of geometry.

This desire for a "satisfying" mathematical structure is something we have already encountered in our earlier discussion of the role aesthetic issues play in the acceptance of a theory in mathematics and physics. Such considerations are entirely capable of trumping the meager deliverances of experience and/or what is strictly derivable by formal means, when these fail to supply the grounds for a sufficiently rich mathematical structure. Thus, in developing the idea of a covariant derivative to be employed in a mathematical space such as is used in GR, it is not enough that it behave in some more-or-less P.D.E.-like manner while at the same time transforming in a properly tensorial fashion. There are, in addition, at least two other desiderata which must be built into the covariant derivative in order that a genuinely "satisfactory" metrical space will be developed. The first of these is that the indices of the covariant derivative should obey certain rules regarding commutation between the indices of the derivative itself, and the object on which it is operating.[75] Secondly, when the function being differentiated covariantly is a "scalar" function[76] the covariant derivative should reduce to regular partial differentiation. In terms of the practical requirements of this kind of geometry, both rules are probably equally important, though the second rule seems more obviously necessary in order for the covariant derivative to maintain its P.D.E-like operation. And yet, as Carroll points out, "[t]here is no way to 'derive' these properties; we are dimply demanding that they be true as part of the definition of a covariant derivative" (Carroll, 96). These properties lead to attractive features in the nature of the covariant derivative, but there is nothing in the situation beyond our practical exigencies to force these properties to be "true" of that operator.

So what have we learned from the preceding? No part of what has been said here, either in this chapter or the prior ones, can or should be construed as a criticism of the general methods and techniques of mathematics or physics. But we are also, quite certainly, compelled to acknowledge that just as there are aesthetic issues at play in the acceptance of mathematical and physical theories, so there are pragmatic ones as

---

75. Specifically, "contractions should commute." The details of this particular mantra need not concern us.

76. I.e., one which is a coordinate independent map from the space in question to the real numbers, rather than itself being dependent on the coordinate system chosen, or else mapping from various points of the space to, say, tensorial or vectorial structures.

well.[77] These are not faults to be done away with, but the essential materials that motivate and energize practitioners in these fields to pursue their inquiries to some kind of relative completion. Inquiry in the absence of aesthetic and pragmatic considerations is not just blind, it is pointless. But as with aesthetic criteria, so with pragmatic ones. The selection of such criteria is not absolutely forced upon us by purely "objective" circumstances; rather, human decisions and preferences are mandatory contributors to the process. Just as in the previous chapter we observed how aesthetic matters played into the development and ultimate acceptance of general relativity, so too here we begin to see how pragmatic considerations are necessarily brought to bear so that the requisite machinery for a theory of cosmology *can* be formulated, making it possible to at once unify a wide array of empirical phenomena while at the same time embedding these phenomena in the mathematical formalisms needed to give numerical outputs from the formal theory that would correspond to the physical actions of measurement.

But note here that while the deeply intertwined ideas of the metrical tensor, connection coefficient, and covariant derivative provide us with the machinery for producing numbers to compare with our measurements, nowhere does the theory tell us *how* we are to know each and every specific metrical $g_{\mu\nu}$ associated with each and every point in space. And while the covariant derivative will give us a numerical value for the rate of change in the curvature of space in the neighborhood of any point in space, it does not give us any explanation for its own possibility, or how we are supposed to know its operational structure at points of space far beyond our direct experience, whose metrical relations are determined by configurations of matter and energy we have no way of gauging absent a complete knowledge of those very metrical relations. In other words, what we have seen here are the structures of a *metrical* theory. We have yet to bring anything like a *theory of measurement* to bear on these problems. For the moment, though, we will continue our focus on metrical theories in section two.

II. Comparing Metrical Theories:

The PPN framework was introduced in the previous chapter as part of the background of the development of general relativity and cosmology. Here we are going to consider some more of the details of PPN itself so as

---

77. These two criteria, the aesthetic and the pragmatic, bleed together so completely that one cannot hope to give a hard and fast distinction between them. But the rough and ready conditions we have discussed here are quite adequate to our purposes.

to further expand our examination of metrical theories in general, and to deepen our appreciation of some of the pragmatic factors that have been at work in the acceptance and rejection of various cosmological hypotheses.

Now, it was stated in the previous chapter that the PPN formalism was rather like a ten parameter "machine" into which alternative theories of spacc and gravity could be plugged and thereby compared, by variously and selectively activating appropriate combinations of those ten parameters. But such a description scarcely does justice to the exceptional work required to put this capability into a single mathematical system. There are, as was mentioned, a great number of such theories in the literature, each of them developed with a considerable level of mathematical finesse. Thus, for example, even though it was among the first alternatives to Einsteinian general relativity ever proposed, Whitehead's own theory proved difficult to incorporate into the early PPN framework. In 1971, even while Clifford Will was using Whitehead's theory as one of his primary examples in his discussion of theories of gravity, he freely admitted that it was "too complicated to fit into the (*then*) nine-parameter PPN formalism" (Will 1971b, 143). It was only with the publication of (Will 1973) that the PPN system was expanded to the current ten parameter structure that could directly accommodate Whitehead's theory with the addition of a specific "Whitehead" term (See also Will 1993a, 98 and 104).

It is perhaps unsurprising that the formalism itself requires a measure of subtlety in its use. To begin with, while there is much to be admired about the truly enormous amount of effort that was required in order to make the PPN formalism work, this is achieved at the cost of transparency. PPN is "a 'super metric theory of gravity' whose special cases (particular values of the parameters) are the post-Newtonian metrics of particular theories of gravity." (Will 1993a, 97.) But this all-encompassing nature necessarily requires that the PPN formalism is inherently more complex than any of the individual theories it is capable of mirroring within itself. Much of that complexity can be peeled away as one moves between particular instances, since these sub-theories will (typically) activate some parameters but not others. Yet one will never reach a point where the PPN machinery is as straight-forward as any of its encompassed sub-theories. For amongst those parameters that are engaged by the particular theory there will often be formal structures whose presence is not required by the particular theory, but must be held as latent possibilities to mirror other particular theories than the one under consideration. However gracefully they might otherwise fulfill their purposes, highly generalized, "all-things-to-all-people" types of systems will ever and inevitably be far more complex than the specialized systems they are intended to imitate.

This leads rather directly to a second factor to consider. There are inevitable biases in how the PPN formalism represents alternative theories of gravity. Thus, for instance, one must decide how to deal with physical theories that contain "absolute" elements or "prior geometries." Will remarks that, "[s]ome authors regard the introduction of absolute elements as a failure of general covariance (Einstein would be an example)" (Will 1993a, 17). Hence, the decision to include theories that have such elements in the PPN machinery can be viewed as at least a little controversial. But, in a certain sense, all such decisions – what to leave in, what to exclude – involve extensive possibilities for controversy. Thus, for example, Will makes the following observation:

> [T]he most general post-Newtonian metric can be found by simply writing down metric terms composed of all possible post-Newtonian functionals of matter variables, each multiplied by an arbitrary coefficient that may depend on the cosmological matching conditions and on other constants, and adding these terms to the Minkowski metric to obtain the physical metric. Unfortunately, there is an infinite number of such functionals, so that in order to obtain a formalism that is both useful and manageable, we must impose some restrictions on the possible terms to be considered, guided in part by a subjective notion of "reasonableness" and in part by evidence obtained from known gravitation theories (Will 1993a, 93).

One is, in fact, faced not only with an infinite range of functionals to choose from in defining the metric structure of a *particular* theory, but also (and, in the context of Will's comment, especially) confronted by the equally broad range of choices as to how to represent such a particular theory within the machinery of the PPN framework.

And, as Will says, part of the criteria of reasonableness is based on factors which are themselves predicated upon "evidence obtained from known gravitation theories." The best known such theory, of course, is GR. Unsurprisingly, GR's representation within the PPN formalism requires only two of the ten parameters, all other factors being set to zero. This makes GR the baseline against which all other theories are compared. In contrast, even though Whitehead's theory is almost as old as Einstein's, it was more than fifty years after its introduction before it was fully incorporated into PPN, involving not only the same two parameters as GR, but the special "Whitehead" parameter created precisely for that purpose and four others as well (Will 1993a, 139). Because of this, not only are alternative theories of gravity rendered at least partially parasitic upon the presupposed *mathematical* structures of GR, rather more importantly for our purposes, they are also (again, at least partially) reinterpreted into the scheme of nature which underlies GR. In other words, they are *philosophically* parasitic as well. Thus, for example, Will in commenting upon the so-called "prior geometry" theories notes that the introduction of such "absolute elements" as a prior geometry is often viewed as a sign of incompleteness in the theory, since such elements "are generally *not* derivable from 'first principles'" (Will 1993a, 18). Yet Will offers no discussion of Whitehead's evidently quite exceptional deployment of just such first principles in his own theory, not even to the extent of noting their existence.

It was, of course, precisely such first principles which motivated Whitehead to introduce his theory in the first place. In addition, one can scarcely resist observing the irony of Whitehead's work being bracketed by Synge previously complaining about too many first principles, and then Will's later complaints (which do not specifically mention Whitehead) about too few.[78] But, as previously mentioned, Will is getting his information on Whitehead's theory via Synge, both directly and through his discussions with Wei-Tou Ni (See Will 1971a, 153 fn's 9 and 10). It is not clear if, or to what extent, Will ever directly engaged Whitehead's full argument. While R is listed in the bibliographies of (Will 1971a) and (Will 1993a), for example, there is no attendant discussion in either of those works to suggest that Will referred to any more than the mathematics in the second half of the book. Nor, for that matter, does Will reference any secondary literature on Whitehead except for works such as Synge's which specifically eschew any attempt to look at Whitehead's deeper arguments.

Finally, we can close this section with a few more comments on the nature of viability of cosmological theories, and the meaning of the term "metric" as this is used in the physics literature. To the first term, clearly a viable theory is one which at once makes substantive and testable predictions about the world, predictions which lend themselves to rigorous testing, and which the theory ultimately passes. Up until 1971, Whitehead's theory met all of these criteria. It was only in 1971 with the publication of

---

78. Although, one should also note in this context that Will does not explain what he means by the term "first principles." Given the expected audience of the book in which that claim was made – primarily physicists and persons with a very strong background in mathematics and science – it is perhaps appropriate to suspect that Will means something quite different from Whitehead's more philosophical sense of the phrase.

Will's work that the explicit mathematical formulation[79] of Whitehead's theory was arguably shown to violate measured results. If this argument holds, and we shall take time to question it in chapter five, this particular aspect of Whitehead's theory can no longer be considered viable. But, as already noted in chapter two, there are other bimetric theories[80] which are still viable, though they seldom if ever receive any notice outside of the specialized physics literature.

So as long as a theory produces numbers that appear to work, which is to say, it delivers numbers which can be compared with with physically achievable measurements, then all appears to be satisfactory. But nothing is said about the theory's underlying claims about *how* those numbers are gained, interpreted, or even presumed to be possible. That the numbers are produced by the theory, and appear to work when compared with measurements, would seem to indicate that they *must* be possible.[81] So, because it would seem that such an account must be possible, the need to find such an account becomes of little direct importance to the practice of physical science, or decisions regarding the comparative viability of any particular theory. But the need for such an account does not go away simply because the theory "works" in some short-term sense of that term. Absent a full account of how measurements are possible, physical science is left in the position of the fellow who did not so much believe in miracles as rely on them. One has an impressive list of numbers, but no real understanding of what those numbers mean.

As to the concept of a "metric" theory of gravity, as previously mentioned this is a different concept from that of "metric*al*" theories at large. At this stage of our science, only "metric" theories of gravity, a characterization that also includes Whitehead's, can possibly be viable accounts of space and gravity. On the other hand, a *metrical* theory, as I have been using this term, is simply one that produces numbers which can be compared against measured observations for the purposes of test. Consequently, non-metric theories of gravity in the sense given this phrase in physics, can be *metrical* in the sense that I have used here.

---

79. A great deal of hay will be made in chapter 5 regarding these calculatedly narrow qualifications.

80. Again, Whiteheadian style theories that have a dual metric structure. Usually, one of these metrics corresponds to what Will has called an "absolute element," typically a "prior" geometry, while the other metric deals with the contingent aspects of physical nature. A great deal more will be said on this subject in later chapters.

81. The "First Law" of metaphysics is, after all, "What is actual is possible."

But a metric theory of gravity makes a very robust claim about the relation(s) between space and gravity, to whit, that space is the manifestation of a metric structure, and this metric structure is an expression of gravity. (See, for instance, Will 1993a, 22 ff). But, as we have already mentioned, Whitehead's theory is predicated on a criticism of any attempt to merge the physics of the universe with the geometry of space. So, how is it that a theory such as Whitehead's can at once be a metric theory (in the physics sense of this term), which evidently means it must assert such a close connection between gravity and the metric structure of space, but at the same time criticize this very connection?

It is here that the *bi*metric structure of theories such as Whitehead's and others becomes vital. GR asserts that there is *only* the gravitational metric field; it is for this reason that the tensor representing this metric structure is *fundamental*; it is the exclusive representor of the connection between gravity and space. Bimetric theories, on the other hand, do not deny the existence of the metric relation between gravity and space – as already mentioned, no viable theory can. What these theories deny is that gravity is *alone* in producing this metric structure. The "absolute element," the "prior geometry" – and, as we shall see later on, both of these phrases are at best unsatisfactory and quite possibly downright misleading in the context of Whitehead's work – is the second metric field, which interacts with the gravitational metric, but is not subject to the purely physical *contingencies* of that latter.

This leads to a point regarding terminology which is symptomatic of how Whitehead's theory has been handled over the years. In the classification of theories which Will employs, Whitehead's theory is not listed as a "bimetric" theory; rather it is classed as a "quasi-linear" one (Will 1993a, 138 – 9). This is because Whitehead's explicit formulation in R attempts "to describe gravity by means of a linear field theory on a flat spacetime background." (*ibid.*) Recall, that "linear" means that the mathematics behaves in some substantially number-like manner. On the other hand, the "flat" in the above means that the spacetime metric behaves in some suitably "Euclidean" fashion.[82] There are two things to note here.

---

82. Although, it need not simply *be* Euclidean. There are structures such as cylinders and torii which, to the casual glance, certainly appear to be curved. However, these structures (and others not so easily visualized) can be "unwrapped" to form "flat" sheets, all of whose essential relations can then be translated directly into Euclidean relations. Thus "flat" encompasses a broader range of geometric forms than the purely Euclidean, but it does not' reach into the essentially curved relations which we find in GR and other mathematical theories.

The first point, which we can only mention now but will examine in more detail in later chapters, is that because Will has identified the connection with a "flat spacetime background" as the primary source of the non-viability of Whitehead's theory, Will's discussion fundamentally *fails* to serve as a disconfirmation of Whitehead's *general* theory, which *explicitly* rejects any essential commitment to the flat metric.

But the second thing, and the one we must be clear on now, is that Whitehead's theory should in fact be classified as a "bimetric" one. We can see, for example, that in his responses to Temple's generalization of his theory, Whitehead was very clear that Einsteinian relativity "recognizes only one field of natural relations," while his, on the other hand,

> postulates *two* fields of natural relations, one of these (viz. space and time relations) being isotropic, universally uniform, and not conditioned by physical circumstances; the other comprising the physical relations expressed by laws of nature, which are admittedly contingent (Temple, 193, my emphasis).

Whitehead felt no reservations whatsoever in expressing his thorough respect and admiration for the quality of Temple's work in generalizing his theory into spaces of constant curvature. However, Whitehead was immediately and emphatically clear that the essential component of his theory was its *bi*metric structure. The division between the essentially necessary and "prior" geometrical structures on the one hand, and the purely contingent physical relations on the other was not optional. Specializing Whitehead's theory to the "quasi-linear" category is accurate only in so far as one is content to ignore the larger program Whitehead intended his specialized theory to serve. Consequently, we will continue to resist the accepted usage within the physics literature, and refer to Whitehead's theory as a bimetric one.

To conclude this section, while it is important to understand that the physics literature uses this term "metric" differently from how I have been using, and will continue to use the term "metrical," the distinction is not one that will serve us here. I will not have any reason to ever refer to what a physicist would call a "non-metric" theory of gravity anywhere else in this investigation. Consequently, throughout the remainder of this work, any time I use either the term "metric" or "metrical" it will be exclusively in the broad sense in which I have been using it, to mean anything, whether in theory or in practice, which yields a number which can be compared to, or is a product of, observed measurement.

III. What is Measurement?

So, what we have seen here so far is that any viable physical theory of space and gravity is going to endow space with a robust metrical structure which will itself have an intimate connection with gravity, expressed mathematically in the form of systems of tensorial relations. Exactly how intimate the connection is between space and gravity remains a matter of some theoretical dispute, with different theories expressing differing kinds and depths of connection between the two. For GR, the connection is absolute; space and gravity simply are the mutual expressions the one of the other. The absoluteness of this connection manifests itself in the collapsing together of geometry and physics. For other theories the nature of this relationship is different, and in the case of bimetric theories such as Whitehead's, there are two interacting metrical fields. One of these fields is an expression of the contingent factors of the purely physical relations involved, while the other represents a set of relations that can be seen as playing the role of the purely geometric, and essentially necessary, relations of space. For these latter theories (though certainly not for all alternative theories to GR) physics and geometry remain separate.

For any of these theories to generate a measurement, one must simply interrogate any given point in space for its physical relations, plug these into the metrical relations for that point, including any special factors relating to either the local or global curvature of space at that point, and the theory determines what numerical values should be the result. One can then take these numerical values and compare them directly to the physically appropriate act of measurement involved, and determine the relative success or failure of the theoretical predictions made, the errors and adequacies of the methods of calculation used, the merits of the actual processes of measurement employed, etc. Such is the function of *metrical* theory of physical cosmology.

However, what is left rather vague in all of this is precisely how we are supposed to interrogate a point in space for its metrical properties, particularly when that point is at an inaccessible distance from us. For all but a vanishingly small bubble of space, that pretty much includes the entire universe. The Pioneer and Voyager spacecraft, for instance, have only extended the reach (at least in principle) of our direct physical measurements to the tiniest sliver of space beyond the edge of our own solar system. On the other hand, our earth and near-earth telescopic observations from both the visible, radio, and x-ray frequency ranges of the electromagnetic spectrum, while providing a more "indirect" kind of yardstick by which distant metrical relations can be determined, have nevertheless achieved spectacular levels of sophistication. But these methods, however sophisticated, remain indirect. A telescope does not go out to a distant point in space to collect the metrical information that is *there*, but rather gathers photons *here* from a direction in space, and reveals as much information as it may regarding the energy and frequency of those photons. General relativity also tells us that we know that those photons followed the straightest path possible, a geodesic, through space. But we are only able to "know" this if we already also know that the characterization of space given by GR is, indeed, true. So, while our telescopes can certainly measure a photon *here*, how are we to know anything about the points of space *there*, not only at the photons presumptive origin, but at each of the "theres" along the photon's path which contributed their metrical relations to the space through which that photon traveled?

This is the reason for a theory of measurement. The metrical theory tells us what sorts of numbers we can or might find if we bring our various yardsticks and rulers to any given point of space. This type of theory gives us the machinery to produce numbers, numbers which can be compared with actual measurements. But a theory of measurement *per se* tells us how to make sense of that process of going to get the numbers. And in particular, when there just is no going from "here" to "there" to make a measurement directly, a theory of measurement will give us the logical reasons for believing that we can legitimately – albeit, only conceptually – transfer the numerical results which we have achieved *here* into some meaningful physical relation(s) which exist and are operative *there*.

I have, in chapter one particularly, talked about this distinction as one between a physical theory, which gives us numbers, and a logical theory, which tells us what those numbers mean. This is a rather fast and loose way of expressing things, which can now be tightened up some. Those numbers are, of course, perfectly meaningful *here*. The real issue is, how far afield do we get to cast our net of theory and assume that the results we get from measurements that are, after all, taking place here, are also in fact leading us to correctly interpret physical relations beyond our direct interactions? What manner of relational structure permits us to make those sorts of assumptions? I have been calling a theory of measurement a *logical* theory, but care needs to be exercised with that description. Distinguishing such a theory as "logical" is not, as was previously noted, in contrast to a failure of "logic" as a purely formal criterion in the mathematical theories of physical science. Rather, it is hearkening back to an admittedly archaic sense of logical which now needs to be clarified.

The first thing to note in this context is that a theory of measurement is not, in any sense, a "logical reconstruction" of a given or initial physical theory. Such reconstructions are not without value, although their true usefulness is sometimes not entirely what their initial proponents might have thought. Thus, for example, it is at least arguable whether the logical reconstructions of mathematics that were particularly popular in the first half of the 20th century have succeeded in showing us what mathematics "really" is. On the other hand, the disciplines which evolved out of those early efforts to reconstruct mathematics along a foundational model have been enormously valuable. Our understandings of proof and truth[83] in mathematics, of the kinds of claims which can be formally justified, and what kinds of justifications different kinds of claims admit of, have all been enormously extended by studies which are indisputably of tremendous formal interest in their own right. So it is no part of our business here to dispute the potential value of such reconstructions. But we must also recognize that this is also no part of the business of what I have called the logical theory of measurement.

At the same time, we must understand that simply because a theory of measurement is a "logical" as opposed to "physical" theory, this must not be construed as suggesting that the theory of measurement is in any sense *merely* formal, or somehow devoid of empirical content. We might well wonder, of course, how it could even be possible for a theory of measurement, which is obviously a theory about physical and empirical processes, could ever be construed in a purely or merely formal sense. However, as we shall see in a moment, this has in fact been the case regarding essential elements of any possible theory of measurement. In the meantime, it must be emphatically stated that, despite being characterized as a "logical" theory, it is most certainly the case that a theory of measurement is ineluctably a theory about possible and actual experiences and empirical contents.

So what exactly do we mean by calling a theory of measurement logical, in contrast to the physical theories of cosmology? Well, the archaic sense of "logical" that we are trying to invoke here is one that John Dewey attempted to recover in the early decades of the Twentieth century.[84] What

---

83. Allowing also for the possibility that this latter term should be scare-quoted.

84. With, it must be added, a spectacular lack of success in terms of how his program was received by the larger philosophical community. Nevertheless, the reader is referred to the opening chapters of his *Logic, The Theory of Inquiry*, vol. 12 of The

is serving to characterize the theory of measurement as *logical* is not that it is focused upon the formal relations of propositions, but that it is focused on the formal and at least epistemologically necessary relations of *inquiry*. This fact is vital enough for us to dare a definition at this point:

> A metrical inquiry – i.e., that type of inquiry which is involved in numerical measurements – is an inquiry which specifically terminates in a comparison or set of comparisons of numerically informed empirical observations with an accepted and uniform physical object which is the the standard of measurement employed in the particular process as well as the founding basis of the "numerically informed" quality of those observations.

Thus, while the numerically informed observations which are the product of measurement are certainly *empirical* results of physical activities, the relations which give those results their qualities *as* measurements are still formal and formalizable, and take on relations of necessity not unlike those in mathematized logics of propositions that nevertheless admit of forceful applications in our mundane activities, consumed as these latter are with contingencies.

It is in this respect that the theory of measurement tells us what those numbers – products of physical action informed by physical theory – mean. The particular physical theory tells us what numbers to expect, and generally where to go to find them. But it is the theory of measurement which gives us the logical basis for accepting that the act of comparison between empirical observation and number generating physical standard is justifiable and intelligible.

But there are two steps corresponding, in turn, to two empirical factors to recognize in this process: first there is some physical standard which is the source and foundation of the numbers produced by measurement; secondly there is the act and method of comparison which generates a number from that standard to be associated with the phenomenon measured. The physical standard, while idealized to an extent in so far as it is set aside in a special environment and rarely if ever directly referred to, is nevertheless a physically existing object of which it is at least possible, in principle, to bring into direct comparison with some other object. Examples which immediately come to mind are such things as the official international meter rod in Paris, or other units maintained by NIST in this country. But, in the absence of this possibility of comparison (our second factor above) these idealized objects cease to fulfill any function at

---

Later Works of John Dewey, Southern Illinois University Press, 1991.

all. It is the *possibility of comparison* which lends these objects the very ideality of their status. Failing such a possibility, these objects are nothing more than brute things devoid of any determinable value. Yet, the flip side of this relation must also be given its due. Absent the real physical standard, no act of "comparison" would amount to anything more than vapid flailing, much less an act of *measurement*.

Rhetorical flourishes aside, the above is a partial gloss on ideas central to Whitehead's theory of measurement as these were presented in the third chapter of R. The gloss is only partial because, while Whitehead is certainly clear about the essential "two-sidedness" of the relations between acts of comparison on the one hand and physical structures as the anchor for the act of comparison on the other, (in particular, see R 49 – 50) he also notes that the formal relational structure of the comparison is much more subtle than a simple "equality." Using an example of our own creation, in saying that a yard of cloth is "equal" to three feet, we are scarcely suggesting that there is an *identity* between either the cloth and the three feet, or that the yard of cloth as measured and the standard yardstick maintained by the National Institute of Standards and Technology are the same thing. When we say that A=B, whatever "A" or "B" might be, it is always in respect to some set or constellation of criteria which (usually implicitly) settle not only the standard of comparison, but also its extent. The bolt of cloth is readily distinguishable from both the standard yardstick maintained by NIST and its representative substitute at the fabric store. So, in fact, what we really mean is something more like

$$A=B \rightarrow \gamma$$

where the "$\rightarrow \gamma$" indicates that the equality is only valid in respect of the criteria of the "character class" (Whitehead's phrase) gamma. This "character class" is a list of specifiable characters, as long or as limited as the particular context might require. The individual characters may be related by very narrowly construed structural relations, or connected by a more liberal system along the lines of a Wittgensteinian "family resemblance." But the character *class* forms an essential element in the context of the specific "equals" relation being employed.

However, this class relation only forms one *layer* in the context of the particular equality. In addition, it must also be stipulated that whichever of

$$A=B \rightarrow \gamma$$

or

$$A \neq B \rightarrow \gamma$$

hold, only *one* character from the relevant class applies to *both* A and B, in respect of their equality or inequality, so that the comparison might be of the appropriate type, and hence be legitimate (R 42 – 4). One *can* compare apples and oranges, provided one is not comparing the apples as apples to oranges as oranges. One must have a *single* character employed for both – as fruit, for instance, or roughly spheroid solids – in order to have a meaningful criterion in operation. But at the same time, the specific character chosen must be chosen from a specifiable class of objects – edible foods, geometrical forms – which anchor the particular character selected in a sufficiently detailed context as to give definable meaning to the claim laid between "=" or "≠" so that the claim will have verifiable content.

Reflecting these ideas back into the previous discussion, it is important to see that these ideas of comparison and equality – or inequality, for that matter – apply not only to bolts of cloth but to metric theories of gravity as well. The PPN formalism has built into itself the assumption that *all* theories of gravity which are compared within its machinery share the relevant characteristics. In particular, it is of the very nature of the PPN formalism to take for granted that time and space conform to the RNI assumption, and therefore can be not only adequately represented as strings and collections of infinitesimal points much like those on the abstract real-number line, but that space and time really *are* just such strings and collections. Once such an assumption is made of absolute uniformity of criteria between theories, then the only thing left to be compared are their specific empirical predictions.

As unquestionably vital as these predictions are – we are, after all, talking about physical science – if we permit the matter to just stand there, we've fallen into the same trap as Synge, which we discussed in the previous chapter, and have in a stroke brushed aside all of the "philosophical baggage" of any theory (such as Whitehead's) which is taking on a larger set of issues than just the most efficient and accurate calculation of numerical results. As we have seen in the case of Whitehead's theory, *how* those numerical results are achieved and given meaning is as essential a part in the entire theory as is the mathematical machinery used in generating those predictions. In chapter four we will explore in some detail how dramatically Whitehead's general approach differs from that of any theory which can casually be incorporated into the PPN system. For now it is sufficient to note how completely the methodology of the PPN formalism shoe-horns all of its compared theories into a single procrustean theory of reality and measurement. No part of the PPN theory attempts to compare the underlying *logic*[85] of the various physical theories which it incorporates, or even notices that such matters might possess scientific relevance.

Returning our attention to the general question of measurement, when dealing with the subject of space, the "equals" relation takes on a special character of its own, and with this special character comes a special name: it is called "congruence." Everything that was mentioned above concerning Whitehead's treatment of the general "equals" relation now applies to congruence as well. Indeed, the importance of this relation in the context of any science of space required a substantive expansion on the ideas Whitehead had already deployed, so as to bring into play aspects of how a standard measure might be "projected"[86] into comparison with the spatial phenomenon of interest. Indeed, the special requirements of spatial relations led Whitehead to develop his previous comments to an even greater level of detail (R 45 – 8). The relation of congruence had traditionally been linked to Euclidean ideas which, in their turn, translated into the direct comparisons of "rigid rods," physical standards which could be directly translated from one point of space to another. In this mode of thinking, the standard yardstick maintained by NIST is not *just* a standard which can be *projected* to other points of space according to logically and mathematically rigorous systems of functionally applicable rules. Rather, it is an absolute and unchangeable ideal that relates to each and every point of space with nothing less than complete identity of being. Such is the rigidity of Euclidean rules. In a "merely" projective system of relations such as Whitehead discussed, the standard yardstick held no more than a *functional* relation to the rest of space, though that function was ultimately connected to space by systematically uniform rules. In the pure, "rigid rod" concept of congruence, the standard yardstick's relation was one of an ontologically, even transcendentally, pre-given nature. Projectively and functionally, the yardstick might translate down to a mere inch at one point, or to many miles at another, provided only that suitable rules were in place. In contrast, the "rigid rod" picture was subject to no such functional

---

85. And, to be honest, one can scarcely fault highly talented physicists such as Clifford Will for such lacunæ. After all, the questions that were not asked were of a specifically *philosophical* nature. "The fault, dear Brutus ..."

86. This term is not an accident. Whitehead, who was a published authority on the subject, was applying principles of *projective* geometry in order to found the relation of congruence on a more substantive basis.

mutations. A yard was a yard, an inch was an inch, and all such relations were absolutely fixed in character for all points in space. An appropriate analogy here is the comparison between the "vertical" lines on a globe (the lines of longitude) and those on a piece of graph paper. On the graph paper, the absolute distance between the vertical lines is the same at every point. On the globe, however, the distance between lines of longitude varies according to how far north or south one is on the globe, although functional and *projective* relations between these lines remain constant.

So, in this respect, Whitehead was clearly trying to move beyond a purely, "rigidly" Euclidean interpretation of spatial congruence relations. However it must be said in this respect that, while his attention to congruence relations tended to set Whitehead apart from most other thinkers both then and now, by itself it did not make him unique. A few of his contemporaries also turned a critical eye to an examination of congruence relations in an attempt to elucidate some of the core meanings of the emerging relativity theory, most notably Reichenbach and Weyl. A very quick glance at some of their work on this subject will be useful to us here. But this glance, like many other aspects of our argument, is dogged by a problem for which there is no resolution: we are glancing at their comments in order to see what they do *not* say.[87] Unlike Whitehead, who will draw his formal considerations of congruence back into the "logical" issues of empirical inquiry, the discussions of Weyl and Reichenbach remain fixated on the formal and mathematical aspects of the congruence relation. Formally, there is nothing wrong with what these latter thinkers said. Its just that, from the purely formal perspective, the mathematical inquirer has a more or less complete command of the subject being dealt with. However far beyond any individual or collective understanding those mathematical relations might lie at any given point in the inquiry, the totality of the mathematical space – whether one believes this to be a Platonic existent, or a purely mental construct – is always present to the mathematician. For physical inquirers, on the other hand, space beyond their direct empirical grasp is absolutely beyond them. Unless there is solid reason for believing that *all* of the essential relations of space are already within our grasp, the physical inquirer is simply and utterly lost. There is no way that the physical inquirer can assume the sort of "God's eye view" on the totality of space that is always available to the pure mathematician – however well or poorly that mathematical "God's eye view" is understood.

87. We have already run up against this problem in our critiques a few paragraphs earlier, of Will's failure to address the alternative theory of nature embedded in Whitehead's particular theory of relativity. How is one to "show" this failure?

It is the absence of this understanding that we must – alas! – "fail to discover" in Weyl's and Reichenbach's discussions.

Thus, for example, in Reichenbach's case, his discussion of congruence is intimately connected to his argument against the *a priori* nature of Euclidean geometry. As a result, Reichenbach works to "decenter" the rigid rod theory of congruence from its established place in the theory of space. His method, however, goes to the opposite extreme of that taken by the Euclidean *a priorists*. For these latter, the rigid rod theory of congruence is an established fact of nature, representative of the ontological structure of the real. For Reichenbach, on the other hand, all forms of congruence are entirely a matter of definition (Reichenbach, 17 and 54). To be sure, the "definition" involved is what Reichenbach refers to as a "coordinative" one, in which there is a coordination of the definition with a factually existing entity. Yet simply because this definition is coordinated with a physical entity, we cannot be permitted to let this distract us from the fact that it remains a definition. As such, the congruence relation is not something to be verified or established in experience, so much as it is *stipulated* by the relevant theory (Reichenbach, 54). Consequently, he insists, "[s]pace as such is neither Euclidean nor non-Euclidean, but only a continuous three-dimensional manifold.[88] It becomes Euclidean if a certain definition of congruence is assumed for it," (Reichenbach, 56 – 7).

We need not concern ourselves here with whether there is, or ever was, any real hope of reconciling such a radically positivistic philosophical approach to things with the actual practices of science. What does concern us is to note how thoroughly Reichenbach has distanced himself from any fundamental empirical connection between the congruence relation and the world in which it is taken to be operating. According to him, our definition of congruence is, "projected by us into space, not discovered in it" (Reichenbach, 54). To be sure, a great deal of this discussion takes place in the context of justifying the conceptual liberty of theorists in the composition of their systems. And certainly it is the case that, for Reichenbach, any physical theory must ultimately be answerable to human experience. But the interpretation of that experience, even for Reichenbach, is hardly just a matter of passive reception. And that interpretation has, as we have seen, no ultimate connection with the nature

---

88. It seems very likely that Reichenbach's use of the term "manifold" here is not so much in the mathematical sense, which even by the time he was writing had assumed some very particular characteristics, but in the more Kantian sense of "*manigfaltigheit*."

of space and time as these are "in themselves."

Such a position would not be acceptable to Whitehead. For while he certainly agrees with Reichenbach that there is a *mathematically* unlimited range of choices for the relation of congruence to be employed in any formal theory, more must be achieved to have a meaningful *scientific* theory than just mathematical adequacy. It is certainly the case that Whitehead agrees that there is a definitional aspect to the nature of congruence. "Congruence – and hence, spatial measurement – is defined in terms of the properties of parallelograms and the symmetry of perpendicularity" (R 9). But this definition is not, for Whitehead, something which is then "projected into" nature. Rather, spatial and temporal relations which are projectively congruent to one another are found *in* nature (R 8), and that "in-ness" is not an arbitrary matter. Science, for Whitehead (and certainly, it would seem, for every working scientist), is about *nature*, and not merely about the descriptions which we call "natural." In the sequel to this chapter, we will at last be spending some detailed time with Whitehead's particular theory of nature, an opportunity which will even permit us to see some of the superficial aspects in which Whitehead and Reichenbach might appear to be going in the same direction. For now, it is more to our purpose to emphasize how completely their views of nature ultimately differ. For Whitehead, nature (which, as we shall see, certainly includes our experiences) is a systematic whole, and that systematicity is a found element, not something stipulated or projected onto it.

For Hermann Weyl the congruence relation is certainly dealt with as explicitly as it is in the case of Whitehead or Reichenbach, but his treatment is even more that of a pure mathematician than either of the other two. Weyl's concern with philosophical issues was always real and manifest in his texts, but his approach was at the same time always informed by the assumptions of the science whose foundations he meant to explore. Thus, for example, his treatment of the traditional "rigid rod" theory of congruence was entirely pedagogical in nature, part of a brilliant mathematician's explication of the basics of Euclidean geometry (Weyl 1952, 11 ff). Similarly, when he turns his attention to non-Euclidean forms of geometry, his method remains that of a mathematician, exploring the beauty of his subject from the God's eye perspective of an all seeing survey of the complete space(s) one is examining. All parts and pieces of it are always already there. Even where human intelligence has failed to grasp the totality of the mathematical relations involved, all of those mathematical relations are spread before us, ready *to be* seen, if only we can (e.g., Weyl 1952, 81 ff). Weyl has not the passionate philosophical agenda of Reichenbach's iconoclastic revolt against the Euclidean *a priorists*. Rather, his is the patronly approach of a dedicated teacher explaining without condescension to enthusiastic pupils, those truths of mathematics of which the elect are already aware.

But what Weyl and Reichenbach share, in absolute distinction to Whitehead's argument, is the mathematician's conviction that we have all of the *formal* relations of space already laid before us, *and these are enough*. Certainly it is the case that Reichenbach makes much more of an explicit effort to connect the formal, definitional structure of congruence with the empirical, factual aspects of science. But this connection remains one of a largely *ipse dixit* quality, a projection of the human mind into a space which – ironically, given his stated opposition to Kantian philosophy – is rather a Kantian "thing in itself," an otherwise barely structured *manigfaltigheit* that is beyond our grasp in its true nature.

And let us finally acknowledge that they were not wrong, *per se*, in making this assumption. For, absent some kind of complete grasp of the spatial relations involved, the congruence relation itself becomes inapplicable, perhaps even meaningless. Congruence requires that we be able to translate a comparative extensive quality from one section of a space to another. Such translation requires that we already be in possession of all the relevant connections and relations as would make such a move possible. Mathematicians have just such a complete command of the spaces they are dealing with, even if they have to declare such a connection as an axiom or definition. Or, contrariwise, when such totality of related connection is not available, one is similarly dealing with a space that is either partially or entirely devoid of metrical structure. There is a great deal that can be said about such spaces: an entire branch of mathematics (topology) is devoted to exploring those things. But, whenever one is dealing with systems of spatial relations at least some of which must terminate in numerical, metrical relations, then topology by itself is not enough. In mathematics, the move to a metrical structure is inevitably achieved by the *introduction* of a distance or other type of metrical function, a function which cannot otherwise be "discovered."

Stipulating metrical structure is a fine methodology for mathematics, but in physics something more is needed. Our congruence relation – however contrived, however convoluted, however complexly fabricated – must, to function as a congruence relation at all, be something we can translate to all parts of space to which we mean to apply and test metrical claims. That is something we cannot simply legislate about space, if we mean to be serious about our science. Whatever else might be the case regarding the congruence relation(s) that are built into our theories, they must be applicable to the world in a meaningful way or else our metrical claims regarding that world collapse into nothing more than wishful thinking.

There are, it would seem only two possible ways in which this can occur. The first is through direct, empirical contact, while the second is through some manner of *a priori* generalization. The first method requires the actual carrying of one's congruence relation to every point of space. Something like this is possible on the surface of the Earth. While inconvenient, one could nevertheless carry the standard yardstick out of its protected enclave at NIST headquarters to every point on the planet and thereby determine the "legitimate" measure of a yard.[89] As already stated, however, this method does not help us with the vast universe beyond our solar system, with which we might reasonably doubt we shall ever have much direct contact.

This leaves us the second method, that of some kind of *a priori* generalization. The "some kind of" is an absolutely vital qualification. On the surface, this claim might seem like nothing else than the mathematicians' method of stipulated relations on a space which is presumed to be entirely within their grasp. And, indeed, this is *one* way in which this kind of approach to the problem can be made. But there is another, and Whitehead argued for it. The argument is made in two steps. First, it is claimed, that space and time have an underlying uniformity of relatedness which serves the *purpose* of the *a priori* generalization. It is this uniformity which gives us the instrument to translate congruence to all points of space, even those beyond any possible direct experience. The second step is remarkable: Whitehead makes the radically Jamesian claim that this relatedness is a *content of direct human experience*. It is not, in this argument, genuinely *a priori*. We do not postulate this uniformity, and we certainly do not legislate it. We *find* it in our experience itself.

We will have opportunity in the following chapters to interrogate this second postulate, and see just how convincing it really is, or is not. Let us end this chapter, however, by recognizing a conclusion based upon three facts, which is all but unavoidable:

(1) GR identifies all of geometry with physics. This eliminates

89. Although, Reichenbach would want to argue, this method could not be used to "falsify" our rigid rod congruence relation, only demonstrate its purely definitional character (Reichenbach, pg. 17).

the very *possibility* of the kinds of uniform relations which Whitehead has pointed out are needed in order to apply the congruence relation to those parts of space that are beyond our direct experience. This is because physical relations are contingent, and cannot be known *a priori*.

(2) Congruence is absolutely essential to the very possibility of measurement. In the absence of a congruence relation, spatial comparison of any kind which leads to numerical relations is impossible. No congruence, no measurement.

(3) We of course do not have direct experience of any but the most vanishingly tiny fragment of the cosmos. Our only hope of applying a congruence relation to the universe beyond our direct experience is via some kind of "*a priori*," or at least patently generalizable, connection to that universe.

The conclusion, then, is that GR undercuts the very possibility of any sort of scientific cosmology which would aim at making metrical assertions about the universe beyond our direct experience.

Whitehead's answer was that GR is fundamentally mistaken. Let us now turn to the first round of the details of his theory to see how.

CHAPTER FOUR
The Nature of Nature

> *Now the question is – Has he observed exactly, or, Has he had exact notions elicited in his conceptual experience? In what sense did he observe exactly one chair? ... Suppose we pin him down to one billionth of an inch. Where does the chair end and the rest of things begin?*
>
> Alfred North Whitehead, "Mathematics and the Good."

I. Routes to a Theory of Measurement:

It is a central thesis of this chapter that the theory of measurement and the theory of nature are mutually implicative, the one of the other. However, such a thesis is hardly self-explanatory. If the only tool – or, at least, the only important tool – with which we can distinguish one theory of measurement from another is the underlying theory of congruence, then one scarcely has anything at all with which to make a distinguishing remark regarding the "nature of Nature." For while the congruence relation is of undoubtedly central importance in any theory of measurement, it cannot by itself provide a complete answer to the question, "what is measurement?" True, we presumably have the physical standard that provides that part of the comparison, and we now have at least a sketch of the formal and logical relations needed to bring that physical standard into comparison in a meaningful manner. But there is still the question of the "what" with which that physical standard is being compared. It might seem on the surface at least, that this "what" is, if not obvious in its own right, at least sufficiently independent of the first two to require nothing more than an equally independent treatment of it as it is in its own right. But to make such an assumption is to throw the theory of measurement back upon an *implicit* theory of nature, thus leaving us with a truncated theory of measurement and an uncriticized theory of nature.

Building on the previous chapter's discussion, we can quickly see that the question, of whether the "what" should be included with the relational aspect of the theory of measurement, or somehow incorporated into the nature of the physical standard that is utilized as the anchor of the process, is itself fundamentally erroneous. These are not two independent and utterly self-sufficient factors contributing only externally to the measurement process: rather, each is partly constitutive of the other, and they function as the basis of interpretive possibilities each to each. So any choice of how the question of the "what" is explicated here will largely be a matter of relative emphasis and practical exigency, rather than a commitment to any sort of ontological or categoreal scheme.

A word here on our method is in order. This chapter will be the most overtly exegetical with regard to Whitehead's triptych out of this entire investigation. Our approach here is quite frankly that of an advocate. It is certainly the case that there are aspects of Whitehead's theory that are open to criticism, but we have not the time, nor is it any part of our purpose, to present such criticisms here. Rather, our business is to gloss, within the confining strictures of this single chapter, Whitehead's alternative theory of nature. Whether or not such a theory is "true" is almost immaterial to our task. What we need to establish here is that an alternative theory of nature is at least initially *possible*. If an alternative is possible, then the failure to carefully criticize the standard theory becomes all the more pressing as an issue.[90]

And the failures to critique that dominant paradigm are legion. Thus, for example, to take measurement seriously is to make a commitment to the nature of the thing being measured. In most – if not all – of the physical sciences, that commitment is coupled to the implicit acceptance of the RNI hypothesis, which in turn means that the thing being measured must somehow be comparable to the Real Number line. The RNI hypothesis takes it as given that this comparison is direct and one-to-one – an isomorphism, the "I" in "RNI." This in turn makes enormous claims regarding the structure of the real, which are once again generally left uninterrogated. Is space *really* composed of infinitesimal points? Are the topological characteristics of those points really those of the real number line, at least when taken on a sufficiently local basis? The RNI hypothesis commits us to these ideas, among others. If, in contrast, one were to challenge some or all of these ideas, then it would be necessary to say *what* it is that is being measured such that it is *not* in such a one-to-one relation to the real number line, while at the same time saying why such a comparison is such an effective idealization within natural science. It is only with the realization of this program – or, conversely, the explicit acceptance of the RNI hypothesis – that one will be in possession of a *complete* theory of measurement, a theory which will have been completed by the full development of the theory of nature latent within it.

This is our task in this chapter, to complete our theory of measurement by making fully explicit our theory of nature. This development will require the explicit rejection of the RNI hypothesis on two accounts. In the first but less important place, casual belief in this

---

90. A more detailed discussion of Whitehead's theory of nature and natural science can be found in (Palter 1960).

thesis has contributed to the purely formal and "mathematized" approach to both nature and measurement. Our too casual acceptance of RNI has blinkered us to the questions we ought to have been asking about nature, by providing us with mathematically satisfactory answers we never sought to wonder after because they were just so convenient. This makes possible the purely formal approaches to congruence we saw in the last chapter from people such as Reichenbach and Weyl, a mathematical approach which tends toward a kind of "God's eye view" on the nature of space itself. If and when we do finally wonder after these convenient answers, if at last we confront the RNI hypothesis itself, we discover what is, indeed, the most salient issue here of all: there is nothing about the RNI hypothesis which finds its place in *actual* human experience. The idea that there is a direct, one-to-one correspondence between the structures of space and time, and those of the Real Number line, is a pure abstraction. It is certainly a useful abstraction, but it is one that wants an explanation.

We have been predicating much of our argument here on Whitehead's work in this area, and our discussion on the theory of nature will certainly be no different in this respect. But before turning to the specifics of the theory of nature, as these were especially developed in Whitehead's thought, it will serve to deepen our appreciation of the mutual relationships between the theories of measurement and nature to examine a bit of the history behind Whitehead's own work on these subjects. As always, Whitehead himself provides preciously little assistance in tracking the sources and historical trails of his thought, beyond that which is to be discerned in his published work. So our remarks on this subject will not only be very brief, they will also be highly speculative.

But our speculations are not without foundation. While Whitehead was, of course, a mathematician long before he was a natural philosopher, the one discipline scarcely precludes an interest in the other. And, indeed, the history of Whitehead's published work shows an ongoing concern with the "problem of space." As previously noted, even in his first published work, *Universal Algebra*, Whitehead is focusing a major part of his effort on geometrical problems, including manifolds and Grassmann's "calculus of extension." However, by itself, this and other, later works do not necessarily speak to more than a mathematician's concern with space, at least not until we encounter the material of the triptych and the preliminary essays Whitehead published from a few years prior, where the issues of GR at last made themselves manifest. But, happily for us in this instance, the evidence of Whitehead's interest in the problem of space goes beyond that which is explicitly deposited in his published works. Almost in spite of himself, it would seem, Whitehead left us a trace of that history as it applied to him. A letter from Whitehead to the Provost of University College London, believed destroyed in the German Blitz on London during the 2nd World War, and only rediscovered in 1970, gives Whitehead's own account of the aim and intention of his work at the time of that letter. Dated March 16th 1912 , it is a fascinating insight into the personal life of a man who, especially by 20th century scholarly standards, was particularly careful to exclude his personal and private life from any manner of public scrutiny.[91] In this letter, Whitehead presents his own interpretation of his work to that point. He states,

> During the last twenty-two years I have been engaged in a large scheme of work, involving the logical scrutiny of mathematical symbolism and mathematical ideas. This work had its origin in the study of the mathematical theory of Electromagnetism, and has always had as its ultimate aim the general scrutiny of the relations of matter and space, and the criticism of the various applications of mathematical thought (Lowe 1975, 86, #17 – 20, where this last refers to the numbers with which Lowe indexed the lines of Whitehead's original letter).

Twenty-two years back from 1912, of course, predates the publication of even Whitehead's first major work, his *Universal Algebra*. As we have already noted, it was the Electromagnetic theory of the Englishman James Clerk Maxwell that was primarily responsible for motivating Einstein's work on relativity, both directly – on account of the issues of symmetry that had not been satisfactorily dealt with by the turn of the 20th century – and indirectly through the multiple layers of experimental evidence which included, but were not limited to, the work by Michelson and Morley. Presumably it is obvious that one must be cautious what conclusions one should choose to draw from the above. But surely this much is beyond dispute: Not only was the "problem of space" as a generic mathematical issue a working theme of Whitehead's efforts from the earliest stages of his career, but the problem of the *space of nature* as this is revealed to us by

---

91. It would probably be extravagant to say that an entire book could be written on this letter (unless one was an unapologetic Whitehead enthusiast). At the other end of that spectrum of extravagance, however, is the pain with which we must content ourselves with nothing more than merely mentioning it, and where it may be found (Lowe 1975). In truth, it is an almost inexcusable distraction from the central purposes of this investigation to say so, yet I feel compelled to encourage the reader to look up Lowe's original article which is, among other things, available online from the *Annals of Science*.

the comprehensive structure of Maxwell's theory of electromagnetism, stands explicitly at the fount of inspiration underlying all of Whitehead's main, independent works, until at least the early 1920's.[92] It does not seem like a stretch, then, to say that for Whitehead, the problem of space and the problem of nature were inextricably linked. Should we see it similarly?

Granted the rhetorical nature of this last question, still we need to address it. And in addressing it, the internal compulsion of the relations mentioned seems unavoidable. For, on the one hand, anyone who proposes to engage the problem of space without simultaneously engaging the issues of Maxwell's theory, has decided to work on some other problematic than that defined by empirical science. Such problematics are certainly legitimate philosophical issues – but they are not *our* issues. Ours is a *specifically* scientific problematic. On the other hand, anyone who would even imagine engaging Maxwell's theory of electromagnetism, is already and ineluctably engulfed in the problem of space. And what else is the problem of space, considered scientifically, but a centerpiece of the problem of nature? Finally, in its turn, the question of space, viewed from a scientific and naturalistic perspective, can only be approached from some sufficiently robust theory that will tell us what it means to engage in spatial measurement.

Turning back, then, to the origins of Whitehead's questions, if we take him at his word that the genesis of his own thoughts on the subject was a scientific theory that was at once spatial and naturalistic, then any attempt to shoe-horn him into a *merely* mathematical view of the matters involved must be viewed with more than average levels of careful skepticism. How very much more than such a purely formal approach Whitehead actually took is a matter we shall turn to in a moment. For now, we, in our turn, must be prepared to recognize the "heads/tails" nature of these measurement/nature questions: there is only a single coin here which we are examining. What is the "nature of Nature" such that it will admit of some particular logic of measurement? What does any given logic of measurement tell us about the "nature of Nature," such that we might legitimately expect that method of inquiry to provide us with valid results?

For contrast, let us consider the nature of an Einsteinian universe. We have already seen how the theory of measurement in GR takes its cue from a mathematician's approach to space; all of it is already *there*, all that is required of the formal inquirer is to tease out the technical details. As we

---

92. I would further argue that this problem is one of the central issues of *Process and Reality*. Only with *Adventures of Ideas*, published in 1933, does this central theme fade substantially from the foreground of Whitehead's work.

have seen, that *formal* "thereness" does not by itself serve to answer the *empirical* questions of how we mere mortals are to discover the metrical relations which are "there" at points effectively beyond our reach, and which are intimately determined by contingent physical factors for which we have no ability to give any sort of account. But that formal "thereness" does provide a rather complete *theoretical* picture of what the universe is like, especially given the fully deterministic nature of the mathematics underlying GR. Indeed, because the orthodox (but usually implicit) theory of nature employed within GR and the greater part of physics collapses nature entirely and directly into the mathematical structures used to describe that nature, we can say that the theoretical picture of the universe is so complete, that what we have essentially amounts to a "Parmenidean block." All of reality is already "there" and the appearance of change through time is an illusion due to the deficiencies of human perception.

There is nothing particularly controversial about this assertion. Moreover, if once we accept the above two premises – that the nature of Nature folds entirely into our mathematics, and our mathematics is itself completely deterministic in structure – then the Parmenidean block picture is a necessary consequence of our theory. It is worthwhile to note how, in this regard, by making all of reality – including and especially time itself – part of an ultimately unchanging "block," time has been given an essentially *spatial* characterization: after all, a "block," whether it is in three, four, or more dimensions, is a solidly spatial object. While we should be once again recalling Whitehead's criticisms of the thoroughgoing uses of spatial metaphors in the mathematics of GR, a couple of citations illustrating this point will serve to concretize this claim.

As an opening example, consider the treatment of the subject given most recently by Brian Greene in (Greene 2004). Speaking of the image of time as a flowing river, Greene says that, instead, "Every moment *is*. Under close scrutiny, the flowing river of time more closely resembles a giant block of ice with every moment forever frozen into place" (Greene, 141, original emphasis).[93] The static nature of time in Greene's description could scarcely be more manifest. It is also interesting to note – or, again, to fail to note, as it were – the absence in Greene's text of any discussion of *nature*. Rather, the term (with variations) that he prefers to use is "reality." This might seem like a rather trivial distinction to be making, but in fact a great deal turns upon making at least some kind of distinction here. Nothing significant hinges on Greene's specific use of terms, or course.

93. Not inappropriately, given the nature of what Greene wishes to claim, this quote comes from a chapter entitled "The Frozen River" (Greene, 127 – 42).

What is significant is the absence of a distinction made between natural science and "something else," whatever that something might be. In failing to make that distinction, Greene pushes all that is real into the mathematical framework of physics.

In contrast, Whitehead takes great care to differentiate his project, which he variously characterizes as "pan-physics" or "philosophy of science." This project, he observes, "is solely engaged in determining the most general conceptions which apply to things observed by the senses. *Accordingly it is not even metaphysics*" (R 4, my emphasis). We shall have more to say on the difference Whitehead is drawing here very shortly, when we turn explicitly to his theory of nature. What is worth recognizing right now is that the focus of Whitehead's work in *The Principle of Relativity* is more restricted than all of reality, a restriction we do not find in Greene's approach. For his part, Greene does address himself to the matter of human experience and "things observed by the senses." But his interpretation of the situation – which is hardly atypical for a physicist – is that, "*The* overarching lesson that has emerged from scientific inquiry over the last century is that human experience is often a misleading guide to the true nature of reality. Lying just beneath the surface of the everyday *is* a world we'd hardly recognize. ... assessing life through the lens of everyday experience is like gazing at a van Gogh through an empty Coke bottle" (Greene, 5; original emphasis). Greene is, of course, a mathematical physicist with a penchant for exposition, not someone extensively educated in philosophical subtlety. Yet it is precisely this penchant and lack which make him a voice worth noticing. He is delineating a large part of the problem we must engage here.

Another voice to listen to in this regard is Einstein himself. In an appendix added to the fifteenth edition of (Einstein 1961),[94] Einstein presented his "views on the problem of space in general" (Einstein vii). His intention was to argue for a purely relational theory of space, such that the "concept of empty space loses its meaning" (*ibid*). Yet, his method of doing so was predicated upon a focus on relations between physical objects, a position which, as we shall see, causes its own problems. But, even more importantly for us, Einstein explicitly endorses in this appendix the "spatialization" of all concepts in the physical sciences. Einstein states, "it is characteristic of thought in physics, as of thought in natural science generally, that it endeavours in principle to make do with 'space-like' concepts *alone*, and strives to express with their aid all relations having the

---

94. This was added in 1952; the date of the English translation is 1961. The first edition was printed in 1916.

form of laws" (Einstein, 162; original emphasis). These statements come on the heels of a discussion of the "psychological" aspects of the experience of time. One can scarcely resist reading into this a reflection, on Einstein's part, back some thirty years earlier to the public debate with Henri Bergson.[95] Einstein never mentions Bergson by name, so reading any such connection into his claim is purely speculative. In any event, whether the Bergson connection was explicitly in Einstein's mind when he wrote the above or not, it is clear that he has taken what Bergson viewed as a principle vice of his theory and elevated it to the position of a central virtue of all natural science. And this, in turn, simply reiterates our earlier point: if time, as science engages it, is essentially "space-like," then all of time partakes of the "all-there-ness" of space itself. Again, we have the Parmenidean block.

Now, at first blush at least, it hardly seems possible to avoid this in any scientific theory that hopes to compete with GR. After all, even in Whitehead's theory, we have a four-dimensional mathematical manifold which is intended to provide us with a physical theory of space and time. But the problem with such a reading of Whitehead's work is not just that we cannot stop here. Even more importantly, we cannot *begin* here either. Thus, for instance, on this point, Whitehead states unequivocally that, "There can be no true physical science which looks first to mathematics for the provision of a conceptual model" (R 39). Whitehead is absolutely clear that the only way his argument can be grasped in its entirety is if one (1) understands the conceptual and philosophical background of his approach, (2) follows the mathematical specifics of the physical theory he develops, and (3) interprets the formal systems he employs along the non-geometrical lines he suggests. Thus the method of physicists such as Synge and Will looks only at (2) above, the mathematics of Whitehead's physical theory, and cannot (by Whitehead's gauge, at least) ever hope to come to grips what he was arguing for.

In contrast, what we have seen of Whitehead's theory so far is only half of (1) above, Whitehead's criticisms of GR, and why these should be taken seriously. In chapter five we will look more closely at (2) and (3). Our task now is to look at the other half of (1), the positive aspects of Whitehead's theory of nature. To this end, we will complete this section with an overview of how Whitehead characterized nature. In the following two sections, we will look more carefully at two aspects of this theory, the theory of time and durations in section two, and the theory of extension in section three.

---

95. *Duration and Simultaneity*, originally published in French in 1922.

In point of fact, Whitehead's characterization of nature can be easily stated in a manner that is both simple and accurate – and which then, of course, requires a great deal of careful exposition. Nature, according to Whitehead, is, "the system of factors apprehended in sense- awareness. But sense-awareness can only be defined negatively by enumerating what it is not." In particular, "sense-awareness is consciousness minus its apprehension of ideality," this despite the fact that, "it is perfectly possible to hold (as Whitehead himself did) that nature is significant of ideality" (R 20). Furthermore, since, "nature presents itself to us as essentially a becoming," then "nature is a becomingness of events which are mutually significant so as to form a systematic structure." As a particular consequence of the above, "space and time are abstractions from this structure" (R 21).

There are two points in the above to be noticed at once: the unreserved seriousness with which Whitehead takes the deliverances of experience, and the equally unreserved seriousness with which Whitehead takes time. On this second point, observe that what is delivered to us by sense-awareness is the *becomingness* of nature; the natural is given to us by its systematically related event-structure. Moreover, we shall shortly see that events are prior to both space and time. Despite a measure of superficial similarity, we will see in the subsequent section how radically this differs from Einstein's "object oriented" approach to space and time as secondary relations. On the first point, note in particular that there is an ideal element of which nature – as the deliverances of sense-awareness – is *significant*, without that ideality being itself something directly present to sense. This will become of particular importance in section three when we touch on Whitehead's theory of extensive abstraction.

But before we can turn to those more detailed examinations, there are a few more items of Whitehead's theory which we need to mention here. Nature, as Whitehead insists, comes to us as a systematic structure. This systematicity displays itself to us along two primary "dimensions" of experience. The events of nature can come to us as focal, fore-ground characteristics which maintain a functional continuity across a temporal stretch of experience. The awareness of such a focal-point of experience is what Whitehead termed "cognizance by adjective." It is these "adjectival" *characteristics* which maintain their *functional* identity across time, and not substantive "objects" *per se*. However, these adjectival characteristics would have no context within which their functional reality could be cognized were it not for a second type of cognizance. We are not merely aware of an enduring characteristic of redness before us, occupying some patch of space through some stretch of time. We are also aware that that patch of space and stretch of time relate to the rest of space and time in some meaningful way. No matter how focused our attention might be on some particular adjectival characteristic, we are also always aware, in what Whitehead calls "cognizance by relation," of the interconnected system of events in which this particular adjective is embedded (R 18 – 9, 62 – 4). These relations form the necessary background which make possible the adjective's functional foreground position as a durable characteristic being cognitively attended to.

But what is more to the point, this relational background – which is a part of nature and as such a content of sense-awareness – is itself the very fabric of the underlying *uniformity* of nature. There is what Whitehead calls a "uniform significance of events," such that

> Though the character of time and space is not in any sense *a priori*, the essential relatedness of any perceived field of events to all other events requires that this relatedness of all events should conform to the ascertained disclosure derived from the limited field. For we can only know that distant events are spatio-temporally connected with the events immediately perceived by knowing what these relations are. In other words, these relations must possess a systematic uniformity in order that we may know of nature as extending beyond isolated cases subjected to the direct examination of individual perception (R 64).

We encounter this uniformity directly: "The truth is that we have observed something which the classical theory does not explain" (R 49). This leads us to the realization that, "the constitutive character of nature is expressed by 'the contingency of appearance' and 'the uniform significance of events.' these laws express characters of nature disclosed respectively in cognisance by adjective and cognisance by relatedness" (R 65; "cognisance" in the above quote is given with Whitehead's spelling).[96]

---

96. Whitehead is endorsing something that looks very much like a Jamesian "Radical Empiricism." It is difficult to estimate how much influence James's ideas might have exercised on Whitehead at this time. Victor Lowe states that "Whitehead was probably acquainted with James's Psychology and perhaps heard much of the ingenuity of the concept of the specious present from McTaggart and others. And from this perspective it is clear that his early empiricism is more radical than atomistic" (Lowe 1990, 105). However, Whitehead does not mention James in the triptych. James's name does come up in *Science and the Modern World*, where Whitehead refers to him as "that adorable genius (Whitehead 1967, 2). Finally, in *Process and Reality*, Whitehead cites James as one of the persons to whom we was "richly indebted" (PR, xii).

Finally, and to conclude this section, when it is said that Whitehead is taking experience very seriously, it must be understood that this is experience in its fullest sense. Whitehead adamantly opposed the bifurcation of nature and experience that threw psychological time and space on one side of an abyss, and the "true" time and space of scientific "reality" on the other, and in the process fabricated an unsolvable philosophical dilemma. (One can see the well delineated shadows of that bifurcation in the Greene quote.) Nature, Whitehead insisted, is "a totality including individual experiences, so that we must reject the distinction between nature as it really is and experiences of which are purely psychological. Our experiences of the apparent world are nature itself" (R 62). We will begin to see how, in section three, Whitehead's theory of extensive abstraction provides a clue to how science approaches truth(s) about nature. But these truths are themselves always a part of a larger whole. Whitehead is not collapsing all of experience into nature, and his very specific definition of nature makes it clear that, for Whitehead, nature is not all that there is. When we take experience at its fullest, we see that, "nature is an abstraction from something more concrete than itself which must also include imagination, thought and emotion." What distinguishes nature is its "systematic coherency" (R 63).

II. Times, Durations, and Spaces:

In Whitehead's account of things, space and time are not in any respect fundamentally constitutive of reality. Instead, what is fundamental is the relational structure of events, from which space and time can be abstracted as representative of real relations in nature, but of no ultimate reality on their own (R 29, 39, 67 – 9). It is the continuum of events which is fundamentally real, and this continuum is a four-dimensional structure wherein each *individual* event is itself a four-dimensional hyper-volume in which "time is the fourth dimension" (R 29). But we must be extremely cautious here. Time is not the fourth dimension of the global system of *events*, but only of some specifically identified *event*. As these events get nested within one another, time does not emerge as a single linear continuum, but rather as a complex system in its own right. What we get is a multi-threaded, multiply-intersecting system of *times*. As a consequence, our discussion here proceeds in two steps: First, we will look at the structure of time as this is revealed in individual events, which in turn means time as it manifests itself in individual experiences. Secondly, we must look at the multiplicity of times, and how this in turn brings about a multiplicity of spaces.

So, to the first point, Whitehead asks a simple yet absolutely essential question: “Have you ever endeavoured to capture the instantaneous present? It eludes you, because in truth there is no such entity among the crude facts of our experience” (R 56). The very idea of such an instantaneous “point” of time is a mathematical abstraction devoid of any direct correlate in human experience. One might object that this is true except, of course, for that highly intellectualized part of human experience which is mathematical speculation itself. But then again, this is an arena of experience outside of *nature* as Whitehead has characterized it. Mathematical points, of either time or space, are never given to us in our sense-awareness, and as such cannot ever be concrete elements of *nature*. Rather, what we encounter in our experience – and, in this regard, we particularly mean, “what we encounter in nature” – are finite but definite slabs of reality whose extensive relations translate into space and time. These slabs of reality, in their temporal characteristics, are what Whitehead calls “durations.” Each duration is precisely that collection of (partial) events which are related as nature “at the same time.” Thus, he tells us that, “a duration is a definite natural entity. ... A duration is a concrete slab of nature limited by simultaneity which is an essential factor disclosed in sense-awareness” (CN 53). Again, this 'at the same time-ness' is not to be understood as 'at the same instant.' Rather, it is an extended system of nature that 'travels together' through the temporal aspect of the relevant events. Indeed, the very use of the word “time” at this stage is rather prejudicial to the subject, and Whitehead advocates speaking only of “the passage of nature,” until the subject can be adequately developed, particularly as “the passage of nature is exhibited equally in spatial transition as well as in temporal transition” (CN 54). It is of the very 'nature of Nature' that it is always moving on. It is only after cognizance has distilled an appropriately stable set of characteristics which display themselves in the passage of events – as opposed to just the passage of nature – that time in something like its scientific sense comes to be recognizable as a genuine, if abstracted, relation *in* nature. In other words, it is only through a collection of adjectival characters that retain a high degree of functional correlation through the extensional structures of some finite but definite set of events, that time emerges as a real character of nature (CN 55; R 68 – 9).

Now, it turns out that these latter events, in which the adjectival characters have been identified, are themselves essentially only portions of larger events. This brings us to some of the more formal characteristics of durations. Durations have extension, and it is the nature of this extension which endows the particular form of the passage of nature within them their peculiarly temporal, as opposed to spatial character. These extensional relations are structured by their various membership and mereological relations, only some of which we can review here.

Of these relations, one of the most important is that systems of durations relate to one another as "families." Now, it is only *within* a family of durations that the membership relation, or any kind of extensional relation in which one duration partly or entirely overlaps another duration, can be meaningfully applied. The relations between *different* families takes on importance in the discussion of alternative time systems. For the moment, though, we will confine our remarks to durations in a single family. Again, this system, this family, does not contain any "points" as members, although (as we will see shortly) point-like structures can be approximated with arbitrary precision. Because the family does not contain points, *any* intersection of non-disjoint durations – finite or infinite – within the family must be itself another duration. There can be no such thing as a "least" duration. In the opposite direction, the union of any collection of durations within the family is, again, a duration. But, in fact, Whitehead makes a stronger claim than just this: there can, in fact, be no greatest member at all. For any union of durations, it is always possible to find a greater duration than this union (CN 59).[97]

Now if we abstractively focus our attention upon some appropriate property or characteristic of nature, we will discover that smaller and smaller durations, each smaller one contained within a larger in this process, will often times serve to effectively refine the important aspects of this natural property. This converging set of ever smaller durations will never reach a smallest duration, nor will it approach a 'limit' of its convergence; there will always be smaller durations. However, the natural properties of this "abstractive set," as Whitehead calls it, "converges to the ideal of all nature with no temporal extension, namely, to the ideal of all nature at an instant." Whitehead immediately points out that this "is in fact

97. For the record, the two rules – the first allowing for infinite intersections and the other stipulating that there is no greatest member – mean that Whitehead's system of durations is not a topological space. A topological space only specifies that all finite intersections are members of the topology. So in this respect, Whitehead's system is too strong. On the other hand, a topological space always has a greatest member, the entire space itself. So in this respect, Whitehead's system is too weak. Despite these admittedly important differences, the relational structure of durations in a family is certainly 'quasi-topological' in nature. A basic familiarity with elementary point-set topology is certainly a serviceable place to start for anyone interested in the subtleties of Whitehead's theory.

the ideal of a nonentity." However, what is important for science is that, "the *quantitative expressions* of these natural properties *do* converge to limits though the abstractive set does not converge to any limiting duration" (CN 61, my emphasis). The ideal of this convergence is what Whitehead calls a "moment" (CN 62).[98]

This 'natural nonentity' of a moment is the approximation to a point mentioned above. In section three we will return to the nature of abstractive sets and what Whitehead called the method of extensive abstraction in more detail. For now, it is enough to recognize that real, natural durations, can be approximated with any finite degree of accuracy using Whitehead's 'quasi-topological' techniques (see footnote 97), while the highly intellectualized, mathematical representations of the scientifically significant content of those durations can be driven to a mathematical limit. Each individual limit, in turn, can be brought into mathematical relation with others. When all of these limits are gathered together, the structure one has is not to be found in *nature*, but is certainly recognizable for all of that: it is, of course, the real number line. This real number system is one of those idealities which are not *in* nature, but of which nature itself *is* significant, and which in turn is significant *for* nature. Indeed, an examination of any history of mathematics will make it perfectly obvious that the real number line came into the universe of mathematical discourse as that structure which completed a systematic array of formal limits that were mathematically interesting. This idealized mirror of nature, in which the last drop of real passage has been wrung out but where serialized order has been elevated to its own 'limit' of perfection, is the ultimate meaning of the concept of "time as a series" (CN 64).

In this respect, each a moment of time is a single point on the idealized formal structure of "time as a series," time as isomorphic to the real number line. However, it is also the case that each of these points is a complete, three-dimensional snap-shot of the spatial structure of nature at that moment. As such, these points have a considerable structure once one takes into account something more than just the idealized extensional qualities of durations. But even here, the structure of these spaces as *spatial* is not independent of the structure of time, if we broaden our investigation now to consider the global structure of time*s*. As already

---

98. But this is not Whitehead's only definition of a moment. He has a second one, which we will encounter in our discussion of the structure of alternative time systems. That second definition might be thought of as the 'external' definition, while this first one is the 'internal' definition, since the first is from the outside while the second is form the inside of the family structure.

hinted at above, Whitehead had a second, 'external' method for reaching the idealized limit of a moment of time. This second definition requires that we now step out of our discussion of those relations common to a *single* family of durations, and take full account of the global *relationality* of spaces and times as these emerge from the full structure of events.

Now, one thing that may have been noticed in the above is that there was little enough in the preceding discussion to yet convince us that there were any important reasons for demurring on the use of the word "time" in favor of Whitehead's phrase the "passage of nature" from the very beginning of our discussion. And, indeed, from the narrow perspective of a *single family* of durations and their associated events, such reasons can scarcely be found. But it is central to Whitehead's argument that there is *not* a single family of durations in nature, which means there can be no single, univocal meaning for the notion of time. We need, then, to ask, why we should think there is more than a single family of durations and what kinds of relations might constitute one collection of durations as a family in contradistinction to another collection? It turns out that the key to understanding this separation of families is in the nature of rest and motion.

Let us recall for a moment that in GR, relative rest and uniform motion are of central importance in establishing a univocal "Lorentz frame" in which the rules of Euclidean geometry are at least locally valid. But this rest / motion is in respect to enduring, substantive objects in an ontologically real four-dimensional spacetime. This is not the same as Whitehead's four-dimensional system of events. To begin with, the notion – presented by, among others, Einstein – that "our notions of space merely arise from our endeavors to express the relations" of enduring objects to one another is something which Whitehead flatly rejects (R 53). On the other hand, "Rest and motion are ultimate data of observation, and permanent space is the way of expressing the connections of these data" (R 70). Taken by themselves, these statements might confuse one into thinking that Whitehead is falling back on a Newtonian notion of absolute space. But nothing could be further from the truth. "Permanent space" is not permanent for the universe writ large, but only for a single, univocal time system in which *that* rest and *that* motion can possess the meanings that they do (R 69 – 71). A particular system of rest and/or a particular system of motion is not a quality of enduring physical objects moving through or resting in absolute space. "[I]t is essential to note that the spatial relations between apparent bodies only arise mediately through their implication in events" (R 71). Specifically, particular systems of rest or motion characterize a particular time system, a close analog to what in GR would be thought of as a frame of reference.

However, rest and motion are relative to a time system and its associated space. As soon as one engages with an alternative field of rest and motion, one has entered the demesne of an alternative time system. This alternative system is internally related by its own family of durations, wherein stable spatial and temporal relations emerge from the four-dimensional relational matrix of events. But note that this latter matrix is not itself space, nor is it time. Space and time emerge from this relational system once a time system is perspectivally engaged by some observer. But they emerge not as an ontological reality but a system of uniform *significances*. As Whitehead points out,

> The paradoxes of relativity arise from the fact that we have not noticed that when we change our time-system we change the *meaning* of time, the *meaning* of space and the *meaning* of points of space (conceived as permanent) (R 56, my emphasis).

But there are no actual – as in, 'ontologically real' – points of space than there are actual points of time. These are only idealized significances which have been teased out of the framework of events by the process of abstraction.

These alternative time systems do not 'live on their own.' Nature, it is to be recalled, is a systematic unity, and each alternative time system is a part of that unity, as a representative of a potential system of spatial and temporal meanings within the overall manifold of events. In particular, alternative time systems will intersect one another, and the structure of that intersection will be a series of moments. Which is to say, the idealized spaces of nature at an instant, within any specific family of durations, will be selected out of the passage of nature and serially ordered by the intersection of that family of durations with another one – the crosscutting of alternative systems of rest and motion will pick out the systematic structure of the underlying matrix of events in such a way as to provide the uniformity of relational structure which, as we shall see below, is necessary to make measurement itself possible.

Consider the figure at the left, which is a projection of some of the four dimensional relationships of events onto the two dimensions of this page. The lines are the idealized three-dimensional spaces of different families of times, the "A" and the "B" families, taken at different

Figure 1

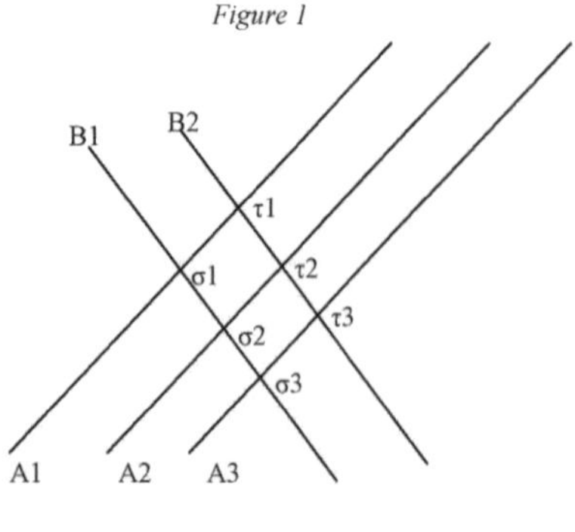

moments, 1 – 3 for the "A" family, 1 and 2 for the "B" family. The interior 'points' of intersection are labeled with a "σ" for those in the B1 durational slab and "τ" for those in B2, while the numbers follow the pattern of the "A" family moments.

Now, each 'line' is actually a three-dimensional space, taken at the labeled idealized moment, while each 'point' is really a two dimensional plane in the respective spaces. Consider now how these two families of times bring structure to each other's spaces. For instance, the interior 'points' labeled σ1 and τ1 represent simultaneous parallel planes within the momentary space of A1. By a similar token, the interior 'points' labeled τ1, τ2, and τ3 are three parallel but simultaneous planes in the momentary three-dimensional space of B2. Taken from the opposite point of view, however, the 'moment' A1 is defined as the particular idealized three-dimensional space it is precisely by its common parallel elements as these are determined by the alternative time system*s* – emphatically plural – which intersect the 'A' family in just the structurally consistent manners that they do. Each time system overlays its own serial structure upon the "permanent" space of any other time system that it intersects. This overlaying of serial orders establishes planes of parallels, another centrally important fact to which we shall return very shortly. This intersection reflects back on the original time system as well, 'slicing' it at an idealized 'point' of its own serial order. This is the 'external' definition of moment mentioned above (R 54 – 5). There are, of course, vastly more than just two intersecting families of time systems within nature – the number is as unbounded as the number of alternative definitions of motion and rest.

Finally, Whitehead makes the rather interesting observation that, "I am assuming on rather slight evidence that moments of different time-systems always intersect. This hypothesis is the simplest and I know of no phenomena that would be explained by its denial" (R 69). Several things are happening in this quote. On the surface of Whitehead's statement, we see that he is assuming there are no "free-standing" families of time. Indeed, every moment – which is a three-dimensional system of uniform significances wherein rest and motion take on meaning and such that the whole has the character of a space – intersects every other moment. These are, after all, only the distilled modes of significance from the global manifold of events. And it is these intersections which structure the locally defined spaces of any particular time system, giving them the requisite uniformities that make measurement possible. It is very difficult to imagine what such a free standing time system might be.[99] However, another thing

---

99. Although, the region within the event horizon of a black hole comes to mind as one

to notice is that Whitehead is not prepared to simply deny the possibility. If presented with phenomena that would benefit from the hypothesis of a free-standing time system, Whitehead is prepared to modify his theory accordingly.

III Extensive Abstraction and Ideal Limits:

Such is a sketch of Whitehead's theory of time(s) and space(s). Our task now is to provide an equally abbreviated gloss of Whitehead's theory of extension, and the abstractive methods by which the extensional characters of nature can be approached with scientifically adequate levels of quantitative precision. As a first step in this development though, a word needs to be said regarding some new terms. It is, of course, well known that Whitehead was notorious for introducing neologisms, a reputation sealed in his metaphysical work, but a habit well entrenched even within the earlier triptych. But Whitehead was never being gratuitous in his manufacture of new terms. There were always important distinctions that he was trying to bring to the fore. In our case here, the issue is once again the geometric metaphors which continue to dominate physical cosmology. Ever suspicious of the too casual application of spatial ideas to temporal relations, or those of passage in general, Whitehead struggled to divest his own terminology of any connotations that might prejudice the issue in favor of just such rampant spatialization (e.g., see PNK 117).

As a result, Whitehead chose to reserve the standard geometrical terms of "point," "line" and "plane" to refer exclusively to the traditional geometric forms that one would find in an idealized, "time-less" space.[100] This latter is a space which is not just an idealized limit of nature at a moment, it is that spatial aspect of a time system in which the geometrical relations have been entirely divorced from any essential relation to any particular moment of the time system (PNK 130). For those relational characters that did possess such vital connections, Whitehead introduced the respective terms "punct," "rect" and "level." Thus, "a point on a line" for Whitehead, in so far as it refers to any natural structure whatsoever, does so without any reference to the system of durations and idealized moments in the time system for which that space is time-less. In contrast, "puncts on a rect have an order which is derivative from the order of moments in a time-system and which connects the orders of various time-systems" (PNK 119).

---

possibility.

100. In CN and R Whitehead will opt for the term "permanent space" instead of "time-less space."

Once again we must emphasize that both groups of terms represent the products of a high degree of abstraction. Whether we are talking about a purely geometrical line or a temporally infused rect, for instance, these structures, while they represent real characters and idealized significances of nature, are reached by a cognitive process of stripping away from that nature some qualities of extension which are really there. Thus, for example, a three dimensional space at a moment is abstracted from true, durational nature by stripping away the extension of time. Or again, a level is a flat surface sheering across times while a plane is a further abstraction from time-less space so as to eliminate one spatial dimension. (Recall that the temporal component in the latter instance has already been removed.) Similar sorts of elimination will serve to reduce an extended portion of nature to such idealized structures as rects and lines, and finally puncts and that most idealized element, the point.[101]

Since what is real for Whitehead are events, in order for these abstracted elements to have any effective scientific connection to nature they must yet correspond – in however abstract a form – to aspects of that nature. The ideal of an instantaneous point is an example. As an abstracted representative of a reality in nature, it corresponds to what Whitehead called an "event-particle." Whitehead characterizes an event-particle as "a route of approximation ... which is an ideal satisfied by no actual event"[102] (PNK 121). This "route of approximation" mentioned above is Whitehead's theory of extensive abstraction. A property of nature is highlighted for some scientific purpose or other, and a converging system of approximations to that property are brought into play, such that as less relevant aspects of nature involved with this property are isolated and stripped away, numerical meanings can be assigned to this property with ever increasing levels of accuracy. We have, in fact, already encountered this process as applied to time in our discussion above (recall the quote from CN on page 134 above). However, this process of convergence is not just limited to extended durations, space can also be approached from

101. As will be discussed in the appendix, Whitehead will later on, in *Process and Reality*, come to repudiate some of this mathematical ontology. We will continue to discuss the earlier theory in the body of this chapter, because that theory is a little more intuitive (especially for non-mathematicians), it is the argument that is found in the triptych, and the greater sophistication of the later theory in no way invalidates the overall argument.

102. For the record, the ellipsis in the above eliminates some terminology which Whitehead would come to discuss and characterize in a very different way once he turned fully to doing metaphysics. By dropping that term out I believe some potential confusion can be avoided without losing the gist of Whitehead's argument.

different abstractive perspectives. Indeed, it is not so much a matter of applying the method to space and or time; primarily the method of extensive abstraction is applied to *events*, because extension is itself primarily a quality *of* events. Space and time are themselves derivative relational structures, and to the extent that they display extensive characters, they inherit these from the system of events from which they are distilled (PNK 61 – 2; 74 – 7). Extensive abstraction is a means for creating ideal simplicity in the relations of events, which can in turn pass on some of that ideal simplicity to the relations of space and time.

Indeed, Whitehead describes extensive abstraction as, "a method which in its sphere achieves the same object as does the differential calculus in the region of numerical calculation, namely it converts a process of approximation into an instrument of exact thought" (PNK 76). It is worth noting, however, that even where the limit toward which the approximation is converging is itself univocal, the routes to that limit need not be. As a very simple example, consider the line below with the "*" in the center. Suppose this is the limiting ideal of convergence. Now, in addition, let "( )" and "[ ]" represent two different routes of convergence.

... ([[([[([[([[([*])]]))]]))]] ...

This diagram differs from a popular one to be found in (Kraus 1998, 148) and (Palter 1960, 43). Yet the above serves our purposes here, for it helps emphasize the differences between the two routes. Note, though, that both converge to the same "point," and there is nothing "topologically peculiar" about that convergence, despite the slight (and deliberate) unevenness between the two.[103] What is important for us is to recognize that there is no reason to suppose that natural knowledge is achieved by any single, univocal path of abstraction.

Where we must turn our attention now is to the nature of extension itself, and how this plays itself out in the theory of measurement. The above tells us how to peel away layers of extension in order to find some numerical value which can be legitimately assigned to some genuine character of nature. But it does not quite tell us how to *choose* those characters. And, in particular, while much has been said regarding the importance of the congruence relation for the theory of measurement, we've yet to know *which* such relation to pick from the literally unbounded

103. And, again, it is the *numerical value* associated with some extensional character or quality of nature, not nature itself which "exists" (in a mathematical sense) as a limit of these two converging systems. Nature itself has no such limit.

range of purely mathematical possibilities. Furthermore, it was Whitehead's contention that we cannot *just* pick a rule of congruence without founding our entire theory of measurement upon a vicious circle (R 51). According to Whitehead, congruence is at once both an aspect of and evidence for the uniformity of nature. "Congruence is founded upon the notion of repetition, namely in some sense congruent geometric elements repeat each other. Repetition embodies the principle of uniformity" (PNK 141). This is so because a congruence relation is, in essence, our ability to repeat the steps whereby our standard unit of measure is brought into comparison with those structures we are investigating. Whether we are laying our yardstick end- to-end to measure the span of a length which exceeds our unit in size, or we are moving the standard in some other stepwise fashion, each one of these steps requires some manner of repetition which is itself an expression of the uniformity of space. This repeatability of congruence is not a mere factor of our definitions, but a "substantial law of nature" (PNK 142). As a result, the notion of congruence is one that must not only be built up from elements that are logically more primitive, these elements must themselves have an objective basis in nature. What is needed, then, is a discoverable set of rules – not necessarily Euclidean in nature – that will enable us to establish the characteristic form(s) of repeatability from which our congruence relations can be derived.

If we return for a moment to the analogy of the actual yardstick, we can picture the tracks that would be drawn through space by the opposite ends of the yardstick as it is moved through space. If the yardstick is moved in a straight line, without warping or twisting the stick itself, these tracks will be parallel to one another along the entire path of the yardstick's movement. At the same time, each pair of opposing points along these tracks will be congruent to one another by the standard of the yardstick. Our view of this process can be inverted, however, so that we do not take the built in congruence relation of the yardstick as basic. Rather, suppose that what we are given are systems of *parallel tracks*. Then these systems of parallels will define congruence for us, by stipulating the uniform projective relations that will determine when two extensions of space are 'the same.' It is here that Whitehead's 'multi-threaded' theory of *times* comes back into play. For, as we saw above, it is precisely the serial structure of time which, when one family of durations intersects another, lays down in the other family's instantaneous spaces systems of parallels. Every time's space is densely criss-crossed by such parallel tracks from the indefinite number of alternative time systems which intersect it. Thus that space is thoroughly 'parallelized' by these time systems, and is thus so richly structured that a congruence relation is derivable from the intrinsic uniformities of the entire system of times and spaces.

Thus, using only his arguments regarding extension as a fundamental structure of events, and his radical empiricist approach to the objective uniformities underlying nature, Whitehead is able to derive the relation of congruence as a discoverable feature of nature that provides the necessary foundations for a physically meaningful theory of measurement. Congruence, in Whitehead's theory, is *not* assumed to be fundamental, and his derivation of it from the structural features of nature is intended to avoid the vicious circularity inherent in any such assumption. What *is* fundamental is extension and the structure of time, as this is derived from the extensional relations of events, and how these manifest themselves in nature. And because these extensive relations of time are in *nature*, they are things which we are aware of. We perceive a character or quality of nature as extending through time, registering both its extended temporality and its uniformity through that extension (as just the character or quality that it is). This relational combination of (1) the perception[104] of (2) an extended duration is what Whitehead called “cogredience” (CN 109). It is important that we understand that cogredience is a relational whole. Let us grant that both perceptions and temporal extensions are – in a sense, at least – independent of each other (although, it must be emphasized, perception can only occur as an act which is itself temporally extended). Yet cogredience itself is more than just the simple sum of these two (PNK 128; CN 110 ff). What is added to the mix is the encounter with the uniformity of nature.

Such encounters are not infallible (PNK 56), and, as we have already seen, neither are they in any sense *a priori*. But the data is *there*, if we are clever enough to distill it from nature The relations of extension and cogredience make it possible for us to discover the natural parallels imposed upon a space by alternative time systems, which in turn makes it possible for us to uncover the congruence relation intrinsic in that space, which finally makes a coherent theory of measurement possible.

It is worth recalling at this point the discussion in the previous chapter of the arguments presented by Reichenbach and Weyl regarding congruence, so as to make as vivid as possible the difference between their

---

104. Whitehead's term was a “percipient event.” But, while it might invite some misunderstanding to use “perception” here rather than Whitehead's own term, for our purposes explaining the difference would only take us on an unnecessary tangent. Whitehead's discussion can be found at (CN 107).

approaches and that of Whitehead. Thus, as already noted, Reichenbach acknowledges the importance of the congruence relation, but sees the selection of some one or collection of them as essentially a free act of intellect governed only by its convenience, scientific utility, and other such criteria (Chapter Three, 115 ff). Weyl tackled the subject like the mathematician he was, noting the mathematical issues, but nowhere offering a constructive account of where the "true" congruence relation might come from. Indeed, both writers essentially treat the congruence relation as primitive, and in this respect show their apparently uncritical acceptance of the RNI hypothesis. Such would seem to be the inevitable result whenever one's theory is driven by mathematical considerations, rather than "logical" ones (in the nuanced sense of that term mentioned in the previous chapter). It is no criticism of mathematics to note that the idealized entities with which it deals, such as points and real number lines, have no corresponding realities in our direct experience of nature. But we *must* have an account of how these ideas and idealized entities are properly applied to nature if ever we are to have an adequate philosophy of the latter. Such an account is not readily evident or apparently forth coming from many authors, including those contemporaries of Whitehead who did at least treat the congruence relation seriously.

It is precisely this which makes Whitehead's theory so worthy of serious attention: he *does* give us just such an account. Whitehead's account is thoroughly rooted in experience, and provides us with the essential tools and systems of relations from which a detailed theory of nature can be discerned, which in turn admits of a coherent theory of measurement that is built out of fundamental relations discovered in experience. Whitehead's discussion in PNK lays out the formal details of these relations at far greater length than we can give full justice to here, while each of CN and R explores these matters with some care. What we need to recognize is that Whitehead's triptych throws down a phenomenological gauntlet, as it were. He presents us with the challenge of, first, looking into our own experience and confirming for ourselves that the structures and relations he has identified are indeed there. Secondly – and this is a matter we will return to in the addendum to this chapter – he invites us to see for ourselves if, indeed, the systems and relations he has set out are not, in fact, logically primitive elements of our experience.

To conclude, then, the primary relations Whitehead has set out for us are extension and cogredience. These are found within nature, which is itself the systematic deliverances of sense-awareness, including the durational characters of those deliverances and the direct relational deposits of those characters in that awareness. Because of the extensional characteristics of events and the cogredient relations of some of those characters as enduring through the spatial structures of our perceptions, the uniformities of space are revealed as aspects of nature from which congruence and finally measurement can be meaningfully derived. The full explication of the details of this theory lies beyond the scope of any single chapter. But what is hopefully made plausible by the above is that the orthodox model of nature which we find operational throughout the physical sciences, and which takes the RNI hypothesis as the fundamental image of the nature of Nature, is not the only possible theory of nature available to us. Whitehead gives us an alternative which nevertheless makes intelligible the success and effectiveness of natural science as this is currently practiced, without trivializing the fundamental connection of experience and nature.

IV. Contemporary Developments:

A few words need to be added regarding the historical development of Whitehead's theory of extension. For, on the one hand, the structure of this theory came to be modified on some important points by Whitehead a few years later in *Process and Reality*, largely in response to some observations and criticisms published in the early 1920's addressing the theory in the triptych. But, on the other hand, these criticisms and Whitehead's corrections were exclusively addressed to various formal and technical issues underlying Whitehead's theory of extension; they do not in any respect address the fundamental arguments that Whitehead advanced. Indeed, it is quite obvious, given Whitehead's repeated citations of his earlier triptych in *Process and Reality*, that he never repudiated his earlier theory of nature, however much he might have modified the logical details of his theory of extension.[105] Consequently, this part of our discussion is little more than an extended footnote which only contributes some minor historical context to our overall subject matter.

Whitehead came to modify his theory of extensive abstraction in response to a series of three articles published by Theodore de Laguna in the *Journal of Philosophy* between July and August of 1922. While the

105. There have been scholars who have argued that Whitehead did, in fact, repudiate his entire earlier theory. See, for example, (Seaman 1955). Seaman's argument – that Whitehead abandoned his earlier work on the philosophy of nature – is disastrously weakened by the fact that Whitehead specifically and positively cites PNK in the paragraph immediately following the one Seaman cites as justifying his (Seaman's) claim.

first two of de Laguna's articles touched on matters of similar interest to those with which Whitehead dealt (as one might gauge from their titles, “The Nature of Space” I and II), de Laguna only mentions Whitehead's work in passing, primarily to note a few of the differences in their approach (de Laguna 1922a, 394). At least one reason for this is that, as de Laguna himself notes, these first two essays had “lain in manuscript for the last five years” (de Laguna 1922a, 395, footnote 1). In other words, the major part of de Laguna's work had been written two years before the publication of PNK. However, the third article, “Point Line, and Surface as Sets of Solids,” published August 17th, 1922, was a purposely written appendix to the first two articles in which de Laguna presented a formally more elegant development of the ideas of point and line, etc., along the lines of Whitehead's method of extensive abstraction. Whitehead recognized the superiority of de Laguna's presentation, and explicitly adopted it into the machinery of *Process and Reality*. In particular, de Laguna's ideas provided a more efficient means of developing extensive relations from which ideas such as 'whole and part' could be derived, rather than taken as primitive, as they are in the triptych (PR, 287).

Quite aside from the wealth of ideas found in Whitehead's *magnum opus*, the page cited above and those following are interesting insights into Whitehead's personality and habits of thought. It is certainly obvious from these pages that Whitehead is not afraid to criticize himself when, after reflection, he decides an earlier position of his is mistaken or inadequate. This demonstrable commitment to truth, as opposed to just his own arguments, is what gives us the confidence to continue using the materials of the triptych without fussing over whether they are consistent with Whitehead's later, more famous works. Whitehead himself would tell us if they were not – and, indeed, he does so where there is a conflict between the earlier and the later.

Probing the details of these developments would serve no purpose here – we've not developed the details of the earlier work sufficiently to make the difference with the later work meaningful. Moreover, the later work is considerably more subtle than the earlier, and as such would involve that much more effort to adequately describe. It is quite sufficient for our purposes to recognize that details we have not probed were ultimately modified in ways that in no way effect the point of the argument we are making here. That argument remains consistent with Whitehead's global philosophical enterprise, and we need not fret over whether Whitehead's mature metaphysical works might somehow call our work here into question. It does not.

However, there is one additional matter in this regard that does merit mentioning. Again, without troubling ourselves over the technical details, it is worth recognizing that the development of Whitehead's theory of extension did not end with *Process and Reality*. As with the empirical theory of R, Whitehead's logical theory of extension went into a long hiatus of all but total neglect once he moved on from the subject. However, unlike the former theory, the theory of extension has enjoyed a significant renaissance of interest. Rather unexpectedly, Whitehead is now quite commonly and explicitly cited as the founder of modern formal theories of the spatial relations of part and whole – mereotopology and mereogeometry. And these derivations of his work – again, with quite explicit citation – are taken as the basis for research into teaching artificial intelligence systems (robots) to see.

This change began in the early and mid 1980's, when Bowman Clarke published a pair of articles in the *Notre Dame Journal of Formal Logic*, developing, modernizing, and (in places) correcting Whitehead's ideas (Clarke 1981 and 1985[106]). Clarke was not specifically a logician by trade; most of his publications were related to process *theology*. But his two articles cited above brought this aspect of Whitehead's work to the attention of an audience that would otherwise have probably never found it.[107] Whitehead's philosophical writings can be quite daunting to the uninitiated, and for someone whose interests lay in more purely formal directions, there would scarcely be any motivation in even attempting Whitehead's *magnum opus* in the absence of some previously established reason for doing so.

More recently, Ian Pratt has further developed the formal relations of the theory of extension, while explicitly acknowledging the previous contributions from Clarke and Whitehead. Thus, for instance, Pratt states, "the policy of taking regions as primitive is perhaps most attractive when considering problems involving mereological (part-whole) and other topological notions – that is, *where no metric information is to hand.* Recent interest in "mereotopology", much of it from within the AI

---

106. One warning on this last citation: Clarke's last name is incorrectly entered into the Philosopher's Index on a number of his publications, including this one, as "Clark." The spelling given here is correct.

107. For instance, one of the leading researchers in the formal logic of spatial reasoning, Ian Pratt, once said to me, "I understand what the theory of extension is doing. But what is the *rest* of that book (*Process and Reality*) all about?" This conversation occurred at the 2003 NASSLLI conference at Indiana University, Bloomington.

community, dates from the work of Clarke, following earlier work of Whitehead" (Pratt 1998, 621; my emphasis). Pratt's published works are extensive, and his citations of Clarke and Whitehead as the opening for his own work quite common. References to the literature of artificial intelligence abound in this context (see, for instance, Randell, *et al* 1992). Indeed, as of the date of this writing (March, 2005) a simple search on Google using the terms ("qualitative spatial reasoning" AND "Whitehead") pulled up 90 hits, while ("mereotopology OR mereogeometry" AND "Whitehead") brought up over 150. This is a lively and ongoing area of research.

While this is not of direct relevance to our argument here, it is convincing evidence that Whitehead's theory of extension is an important contribution to the logic of spatial relations. That it should be such a fertile source of ideas for research, even though that research is in an area beyond our own immediate concern, does serve to validate this aspect of his theory of nature and abstraction as a logically sound starting place, at least, for the purposes Whitehead envisioned. While this does not prove that Whitehead's theory of nature and abstraction is true, or even scientifically adequate, it does at least show that its foundation on his theory of extension – a theory which Whitehead himself continued to develop for almost the entirety of his professional life – is logically sound.

CHAPTER FIVE
Nuts and Bolts

> *Mathematics is the most powerful technique for the understanding of pattern, and for the analysis of the relationships of patterns. ... If civilization continues to advance, in the next two thousand years the overwhelming novelty in human thought will be the dominance of mathematical understanding.*
>
> Alfred North Whitehead, "Mathematics and the Good."

I. Symbolizing Relationships:

This chapter will be our most mathematical by far. But our task remains one of translation and interpretation. As a result, our inquiry will not take us so far as to engage in any sort of mathematical deduction, but once again only so far as the examination of, and familiarization with, a few formulae and definitions, with the hope that we might ultimately come to recognize various patterns when they repeat themselves, even when those repetitions manifest themselves in different forms. This first section will be focused on some of the more elementary textual issues confronting us as we examine a few of the mathematical details of Whitehead's last major scientific work. Focusing upon what is known as a "Christoffel symbol," we will be able to quickly compare some of the ways in which this symbol has appeared in the literature. This, in turn, will make it possible to show a formula of considerable importance both for Whitehead and orthodox GR, and develop this formula through some of these presentational combinations opened up by the different tokens for the Christoffel symbols. This will both clarify some of the key aspects of this formula, and build an appreciation for some of the efficiencies – and, let us be clear on this point, defects – of the contemporary style of representation. With regard to these defects, for the moment let us just note that brevity comes at a price: it only brings clarity for those who are initiated into its deeper secrets, and it only brings efficiency where the abbreviated details are themselves appropriately and legitimately compressed or ignored.

Ultimately, the formula we will examine will provide us with an opening into Whitehead's larger program, both by showing some of its common connections with fundamental ideas of GR in general and Einstein in particular, and then giving us a keyhole through which we can peek at where Whitehead saw his theory diverging from these others. This development will come in section two. There we will take this formula, and a couple of others that Whitehead develops in conjunction with it, and present some of the formal methods by which Whitehead explicitly separated the contingent factors of gravity and physical interactions from the necessary relations of geometry which satisfied his uniformity requirements, and were the basis for what we previously discussed as his theory of "cognizance by relation." In order to develop these ideas, we will deliberately step into a trap that can be encountered in Whitehead's theory by too casually assuming that his mathematics is somehow "really" like orthodox GR. In section three, we will return to some of the previous commentators on Whitehead's, including Synge and Will, and see how they and others have in fact artificially narrowed their treatments of Whitehead's thought, and thus given questionable estimations of the viability of Whitehead's alternative theory of relativity. We will close with some comments regarding the current status of other bimetric theories of gravity which, although developed in complete independence from Whitehead's proposal, nevertheless deserve to be classed as "Whiteheadian."

To our first set of issues, then, the contemporary reader approaching the mathematical sections of R is confronted by a multiply problematic situation. For one thing, Whitehead's symbolism, which was scarcely on the cutting edge even in his own time, is genuinely archaic by today's standards. Moreover, beyond the particular scripts employed, Whitehead would also group symbols together in mathematically legitimate manners that are nevertheless not along the lines one finds in current texts, or even those contemporary with Whitehead himself. Mostly this is just a matter of updating one's typography and arranging one's terms. But this is not something which can be pursued blindly, or in a wholesale manner. For, as we shall see in section two, there are areas of Whitehead's thought in which his use of an older form of symbolism actually becomes an essential component in his argument. Hence, if one too gratuitously "modernizes" his formulae, one risks burying distinctions which are of vital importance to his theory.

Yet another problem has to do with points in Whitehead's exposition in which, if he does not simply mis-state himself, his choice of terms is at least highly ambiguous. Where this might tie in to the first problem is that there are times where the nature of his ambiguity might itself reflect a hold over from older, outmoded forms of expression. As Whitehead himself pointed out in his earliest mathematical work, one of the principle tasks of mathematical reasoning was the elimination of such confusions from the symbols being used (Whitehead 1898, 3 – 4). However, it would be a mistake to suppose that Whitehead had succeeded in eliminating all such ambiguities from his own work. Thus, for instance, on page 76 of R,

Whitehead defines what he calls the "quantities" "ω_μ." However, his own usage on this and the following pages makes it seem more reasonable to contemporary mathematicians to speak of these as *vectors* with a scalar product, rather than as "quantities."[108] There is little reason to suppose that Whitehead was expressing himself in a manner that he would have thought inappropriate at the time of his writing, and this ambiguity may be reflective of the usage of such terms at the time. An even bigger issue confronting us, and a problem that is endemic to any and all mathematical texts, is this: the opportunities for misprints and typographical errors are almost unlimited, in no small part because the persons doing the copy editing and proof reading of such manuscripts must know as much about the mathematics as the original author in order to properly judge the accuracy of what they are reading. Thus, for example, we will also find that his initial presentation of the formula he designates as "Einstein's law" on page 86 of R is in fact incorrect: his use of summations has grouped together terms which are actually summed up independently of one another.

Yet another issue that needs to be mentioned has to do with Whitehead's general method of presentation when handling the mathematical portions of his subject matter. We have already quoted, in chapter one, Whitehead's regret at the need to worry philosophers with mathematics, as well as saddling mathematicians with philosophy. Unfortunately for the philosophers, Whitehead does not invest much if any effort in bringing that group of his readers up to speed on this material. Even when he glosses the basics of tensor analysis in part III of R, Whitehead presupposes a very considerable knowledge of mathematics on the part of his readers.[109] This, and several other factors already mentioned above, come together when Whitehead offers us his version of the above mentioned "Einstein's Law." We are confronted there by outmoded symbols, arranged in a non-standard fashion, containing a minor error in its presentation, referring to a formula of Einstein's which we are assumed to find easily recognizable without any specific citation given for us to make a comparison, or a more detailed description that would allow us to directly identify it.

This is the formula which we will have occasion to examine

108. This was brought to my attention by Professors Kocik and Budzban of SIU Carbondale.

109. Perhaps a mathematician would make the same complaint about the philosophical portions of R, in which case, the breezy cheerfulness with which Synge brushes those parts aside becomes more understandable, even if still not quite justifiable.

momentarily. But we need to build up to this formula, both for our own understanding and to solidify our appreciation of the difficulties inherent in translating Whitehead's work into a more contemporary symbolism. And, indeed, the task we are facing here *is* one of translation, much as if we trying to bring an idea expressed in a foreign language into our own dialect. We discussed this problem some in the first chapter; now we must engage it directly. Our starting point for this process will be the "Christoffel symbol."

In at least one of its manifestations, we have in fact already encountered the Christoffel symbol. It is what we have referred to in earlier chapters as the "connection coefficient." It is worth reminding ourselves here what the connection coefficient is, what it does, and how it is defined. As to this latter, the standard definition is:

$$\Gamma^{\sigma}{}_{\mu\nu} = \tfrac{1}{2}g^{\sigma\rho}(\partial_{\mu}g_{\nu\rho} + \partial_{\nu}g_{\rho\mu} - \partial_{\rho}g_{\mu\nu})$$

where the "g" (with indices) is a matrix giving the metrical structure at each point in the space under consideration, "$g_{\nu\rho}$" and "$g^{\nu\rho}$" are inverse matrices of one another, and "∂_{ρ}" is short hand for $\partial/\partial x_{\rho}$, i.e. the partial differentiation of the term being operated on ("$g_{\nu\rho}$," for instance) with respect to the "x_{ρ}" terms. The connection coefficient itself is not a tensor, but in various combinations it can yield tensorial relations which, in turn, can represent structural and or geometrical relations of the space at its various points. Thus, for instance, the Riemann curvature tensor is commonly constructed from the connection coefficient, and this is a true tensor.

Now, as already mentioned, the connection coefficient is a kind of Christoffel symbol – the *second* kind, to be precise. Here, however, our situation becomes rather more complicated, for there are a variety of ways in which the second type of Christoffel symbol appears in the literature, and these varieties show no particular relation to when or where the original text was written, and as such cannot be simply related to historical or geographical aspects of use. Always those uses involve showing indices within curly brackets "{}," but how those indices are arranged is subject to no underlying standard. Using $\Gamma^{\sigma}{}_{\mu\nu}$ from above as our baseline to keep track of the indices, we find $\begin{Bmatrix} \mu & \nu \\ \sigma \end{Bmatrix}$ is used by (Bishop and Goldberg), but (Weyl 1952) used $\begin{Bmatrix} \sigma \\ \mu & \nu \end{Bmatrix}$ instead. Some authors, including Einstein as early as his 1916 paper on General Relativity, preferred to write $\{\mu\nu,\sigma\}$ (Einstein *et al*, 132 ff). This was Whitehead's preference as well, only Whitehead always explicitly wrote out both the matrix term that was involved in the operation (the "g" from above, the "J" immediately below) and the specific point at which the operation was taking place (the superscript "(u)" immediately below). Thus, in Whitehead's notation, a Christoffel symbol of the second type would be "$J\{\mu\nu,\sigma\}^{(u)}$", meaning the matrix J is operated upon along the indices shown within the Christoffel symbol at the point u.

Now these typographic differences do not effect the meaning of the symbol, its mathematical content or its deductive relations. However, it is significant for our purposes to note that, unlike almost everyone else, Whitehead thought it important enough to continually make the matrix term explicit in his formulations. The reason for this is simple, and will become more evident as we proceed: if one leaves the matrix terms implicit, one has created the possibility for serious ambiguities regarding the metrical structure of the space on the one hand and the gravitational aspect on the other, if not effectively collapsing the two together. In Einstein's case, this was seen as a particular advantage of GR, and consequently the use of $g_{\nu\rho}$ as representing the dual aspects of what he considered a single relation was a conscious choice and a virtue. For Whitehead, it was a vice.

We will follow out Whitehead's thought on this matter more carefully in section II. But as a continuation of our current topic regarding issues of translation, we need to go back yet one more step to the Christoffel symbol of the *first* type. Many contemporary texts, such as (Carroll), do not even bother to mention this type of Christoffel symbol, neither do they bother to introduce the bracket as opposed to the capital gamma symbolism nor make any distinction between the Christoffel symbol and the connection coefficient. For the purposes of contemporary GR, such choices are perfectly reasonable, since there is very little in GR that might need or use the distinction between the two types of Christoffel symbol.

Originally, however, the matter was not so settled. As may be recalled from our discussion in chapter three, the development of the connection coefficient proceeded from the attempts to construct a covariant derivative that would maintain its tensorial invariance across changes of coordinate systems, something which the standard partial derivative would not do. It was in this context that Christoffel symbols were developed. Happily for us, the script for the type one symbol is *relatively* standard throughout the literature. It is typically written as:

$$[\mu\nu,\rho] = \tfrac{1}{2}(\partial_\mu g_{\nu\rho} + \partial_\nu g_{\rho\mu} - \partial_\rho g_{\mu\nu}).$$

While this is a fairly standard way of presenting the type one Christoffel symbol, there is another method which links it more closely – and, in some respects at least, more elegantly – with the capital gamma symbol we have seen in connection with the type two Christoffel symbol. We will touch on this briefly at the end of this section.

For now, however, focusing our attention on the bracket symbolism, with the above definition of the type one Christoffel symbol, the type two is typically introduced as:

$$\{\mu\nu,\sigma\} = g^{\sigma\rho}[\mu\nu,\rho],$$

which, as can be readily seen, amounts to the same thing as the capital gamma symbol used for the connection coefficient. Again, following the standard (which is to say non-Whiteheadian) practice, we have left implicit the fact that the Christoffel symbols are really *operators* acting upon some metrical element. In the cases just above, that metrical element is the "g," which has been left implicit in the deeper recesses of the structure of the Christoffel symbols.

It is worth noting here that the "standard" of usage regarding the "$\Gamma^\sigma{}_{\mu\nu}$" symbol is not something that occurred overnight, nor are the deviations from this practice confined to off the beaten track thinkers. Einstein, for instance, employs the capital gamma symbol in his 1916 essay introducing GR, but he definitionally introduces it as

$$\Gamma^\tau{}_{\mu\nu} = -\{\mu\nu,\tau\},$$

the *negative* of the type two Christoffel symbol (Einstein *et al*, 143). D. F. Lawden motivates this choice a little more explicitly by building up to the idea of the connection coefficient through the study of (covariant) differentiation along geodesics, thus paralleling some of the historical development in his mathematical treatment. Lawden does not use the capital gamma at this stage, but his presentation does develop a three index symbol which he writes as "a_{ijk}" and which he uses to definitionally introduce the first Christoffel symbol as "$a_{ijk} = -[ij,k]$." By multiplying the negative of the first Christoffel symbol by the inverse of the metric (the "g" with both indices high) Lawden raises the third index on his symbol, getting the negative of the second Christoffel symbol, thusly:

$$a^k_{ij} = -g^{kr}[ij,r] = -\left\{ \begin{matrix} k \\ i \quad j \end{matrix} \right\}.$$

(Note, again, the different choice for the distribution of indices within the curly brackets.) Clearly Lawden's "a^k_{ij}" corresponds to Einstein's "$\Gamma^\tau_{\mu\nu}$." However, Lawden then goes on to introduce the capital gamma explicitly, which he characterizes as the "metric affinity," only he makes his version correspond to the *positive* second Christoffel symbol (Lawden 2002, 108 – 10).

These are just a few of the translational difficulties we must confront, and so far we have only been examining the variations around the Christoffel symbols. Our problems become significantly more difficult when we broaden our attention to examine an entire formula. In this situation, of course, the combinatorial possibilities have been greatly increased, and yet we would still like to see at least some part of the meaningful content of that which is being presented to us. A case in point is Whitehead's formula (19) on page 86 of R. Whitehead writes this formula as

$$\boxed{\begin{aligned}&\sum_\rho \left[\frac{\partial}{\partial u_\rho} J\{\mu\nu,\rho\}^{(u)} + J\{\mu\nu,\rho\}^{(u)} \frac{\partial}{\partial u_\rho} \log\left\{-J^{(u)}\right\}^{\frac{1}{2}} \right] \\ &- \sum_\rho \sum_\sigma J\{\mu\sigma,\rho\}^{(u)} J\{\mu\rho,\sigma\}^{(u)} - \frac{\partial^2}{\partial u_\mu \partial u_\nu} \log\left\{-J^{(u)}\right\}^{\frac{1}{2}} = 0\end{aligned}}$$

Whitehead's original formula (19), pg. 86

and tells us simply that it is "Einstein's law" (of gravitation).[110] The diligent reader quite rightly wants to correctly identify this formula with its appropriate correlate in Einstein's work. But even though Whitehead and Einstein were contemporaries and dealing with many of the same ideas using much the same symbolism, nothing precisely like Whitehead's formula above appears anywhere within Einstein's writings.

For one thing, Einstein had already introduced a summation convention where symbols that shared common upper and lower indices automatically had those indices summed when an upper index on one

110. A "J" preceding a Christoffel symbol is metrical tensor, which can be thought of as a matrix, which is in turn differentiated along the appropriate indices by the elements of the Christoffel symbol. The "$J^{(u)}$" means the *determinant* of the matrix J at the point u. This latter is a number, and hence can be meaningfully operated on by the log, the square root, etc. Details regarding determinants can be found in any linear algebra text.

symbol matched the lower index on another. This convention not only simplifies formulae in their full expression, it also makes clear (as we shall see in a moment) the advantage of the capital gamma representation for the connection coefficient over the bracket style of the Christoffel symbols. In addition, when we attend to the summation structure of Whitehead's formula above, we encounter our next problem of translation: the formula, as written, is *wrong*. The error is not massive, and almost certainly due to a mere oversight during composition and proof reading, since Whitehead gives a correct presentation of the formula – (58) on page 185 in part III of R – as part of his sketch of basic tensor analysis. Yet the problem is there in the earlier formula, and in some respects is all the more unfortunate since it appears in the specifically philosophical portion (part I) of Whitehead's book.

The problem is on the first line of the formula, where the summation symbol collects two different terms together within the square brackets. What this literally means is that the "ρ" index in the two different Christoffel symbols within those square brackets are summed over simultaneously, when in fact they should be independent of one another. Still using Whitehead's symbolism, the correct form for his (19) is:

$$\boxed{\begin{aligned}&\sum_{\rho}\frac{\partial}{\partial u_{\rho}}J\{\mu\nu,\rho\}^{(u)}+\sum_{\rho}J\{\mu\nu,\rho\}^{(u)}\frac{\partial}{\partial u_{\rho}}\log\{-J^{(u)}\}^{\frac{1}{2}}\\&-\sum_{\rho}\sum_{\sigma}J\{\mu\sigma,\rho\}^{(u)}J\{\mu\rho,\sigma\}^{(u)}-\frac{\partial^{2}}{\partial u_{\mu}\partial u_{\nu}}\log\{-J^{(u)}\}^{\frac{1}{2}}=0\end{aligned}}$$

Whitehead's formula (19), corrected.

With this correction in place, there are several considerations which can serve to narrow the candidates for the "Einstein's law" that Whitehead spoke of. There are, to begin with, not all that many different formulae involved in GR, especially at the early stages of its development when Whitehead was writing, and there are substantially fewer still of such singular importance that one might feel vindicated in selecting one of them out as specifically *Einstein's* law. And finally, there are only a very few places indeed where a term will show up in the formula with a "log" operator in it, and really only one where such an operator appears twice.

This happens when two of the indices – the one upper and one of the lower – in the Riemann curvature tensor are set equal to each other, or "contracted." As a reminder, the ordinary Riemann tensor is:

$$R^{\rho}_{\sigma\mu\nu} = \partial_{\mu}\Gamma^{\rho}_{\nu\sigma} - \partial_{\nu}\Gamma^{\rho}_{\mu\sigma} + \Gamma^{\rho}_{\mu\lambda}\Gamma^{\lambda}_{\nu\sigma} - \Gamma^{\rho}_{\nu\lambda}\Gamma^{\lambda}_{\mu\sigma}$$

The Riemann curvature tensor

Einstein – who represents the Riemann tensor as "$B^{\rho}{}_{\mu\sigma\tau}$" (Einstein *et al*, 141) – contracts (again, sets equal to one another) the upper index, his "ρ," and the lower right most index, his "τ." Referring to our formula for the Riemann tensor, this would amount to converting the $R^{\rho}{}_{\sigma\mu\nu}$ to $R^{\rho}{}_{\sigma\mu\rho}$, then carrying the indices consistently through in the equation on the right. In Einstein's notation, he is inviting us to examine a new tensor, $G_{\mu\nu} = B^{\rho}{}_{\mu\nu\rho}$. This new tensor has four terms, which Einstein collects into two sub-tensors of two terms each. The first of these (his $R_{\mu\nu}$) is composed of two type two Christoffel symbols, while the second (his $S_{\mu\nu}$) is composed of two "log" terms. If we convert Einstein's Christoffel symbols to the more modern capital gamma, using Einstein's own definition (see above, page 154), then $G_{\mu\nu} = R_{\mu\nu} + S_{\mu\nu}$ becomes:

$$\frac{\partial \Gamma^{\alpha}_{\mu\nu}}{\partial x_{\alpha}} + \Gamma^{\beta}_{\mu\alpha}\Gamma^{\alpha}_{\nu\beta} + \frac{\partial^{2}\log\sqrt{-g}}{\partial x_{\mu}\partial x_{\nu}} + \Gamma^{\alpha}_{\mu\nu}\frac{\partial \log\sqrt{-g}}{\partial x_{\alpha}}$$

Einstein's formula (44)

where the first two terms are "$R_{\mu\nu}$" and the second two are "$S_{\mu\nu}$" (Einstein *et al*, 142). If we do the same conversions on Whitehead's formula (19), dropping his explicit summation symbols (the summation being assumed according to the previously mentioned convention) then we have:

$$-\frac{\Gamma^{\rho}_{\mu\nu}}{\partial u_{\rho}} - \Gamma^{\rho}_{\mu\nu}\frac{\partial \log\sqrt{-J}}{\partial u_{\rho}} - \Gamma^{\rho}_{\mu\sigma}\Gamma^{\sigma}_{\nu\rho} - \frac{\partial^{2}\log\sqrt{-J}}{\partial u_{\mu}\partial u_{\nu}} = 0$$

Whitehead's formula 19, modernized

Other than a trivial rearrangement of terms, this is obviously Einstein's formula multiplied by -1 and set equal to zero. It is also the case that the capital gammas in Whitehead's formulation are all referring to his "J" metrical element, while in Einstein's these are all referring to the "g."

There are a few other points worth noting here. When the frame of reference is chosen so as to make $\sqrt{-g} = 1$, then both "log" terms vanish. With such a reference frame chosen, the equations for the "matter free field" (i.e., empty space) become:

$$\frac{\partial \Gamma^{\alpha}_{\mu\nu}}{\partial x_{\alpha}} + \Gamma^{\alpha}_{\mu\beta}\Gamma^{\beta}_{\nu\alpha} = 0$$
$$\sqrt{-g} = 1$$

Einstein's formulae (47)

(Einstein *et al*, 144). This formula is an essential component in Einstein's development of GR, and has some legitimate claim to the title "Einstein's Law." Also, on yet another notational matter, the careful reader will observe a slight difference in super- and subscripts between Einstein's formulae (47) and (44) above. This is is a matter of no consequence, since each term is independent of the others in the formula, the indices do not refer to any term not directly operated on by the differentials or the gammas. As long as the indices are consistent to the above patterns *within* each term, any arbitrary set of indices will carry the same meaning for the entire formula.

Hence, we can see that Whitehead's decision to set his formula equal to zero is perfectly valid, once one has made the appropriate choices for the reference frame. And his formula coming out as the negative of Einstein's is merely an artifact of his method of derivation. An examination of Whitehead's presentation of the Riemann curvature tensor (R 184) with his second version of "Einstein's Law, his formula (58), shows that he is contracting a different pair of indices than Einstein in producing his version of the formula. In other words, if Einstein's contraction looks like "$R^{\rho}_{\sigma\mu\rho}$", then Whitehead's would be "$R^{\rho}_{\sigma\rho\nu}$". Because of the nature of the Riemann tensor, this choice has the effect of making the one contraction the negative of the other. But otherwise, it is a difference that makes no difference, particularly since the formula equals zero.

Returning for a moment to the type one Christoffel symbols, while it is not often done so, it is also possible to express these in a way that links them directly with the capital gamma used in the connection coefficient. Recalling the relationship between the two types, we can write the connection coefficient as "$\Gamma^{\sigma}_{\mu\nu} = g^{\sigma\rho}[\mu\nu,\rho]$." Now, the "$g^{\sigma\rho}$" on the right of the equal sign is, in effect, simply raising the rho index within the bracket, and renaming it as sigma. We can reverse this process thusly, and get:

$$g_{\sigma\rho}g^{\sigma\rho}[\mu\nu,\rho] = [\mu\nu,\rho] = g_{\sigma\rho}\Gamma^{\sigma}_{\mu\nu} = \Gamma_{\rho\mu\nu}.$$

By multiplying both sides of the expression for the type two Christoffel symbol with the inverse matrix for "$g^{\sigma\rho}$" – i.e., by the *standard metrical tensor of general relativity* – we are left with the original Christoffel symbol of the first type on the one hand, and a capital gamma symbol with all three indices lowered on the other. We have now converted both types of Christoffel symbol to the straight-forwardly tensor-like gamma symbolism. The relation between the two types of symbol is one of raising or lowering a single index using the metrical tensor or its inverse.

Finally, let us see a little of how the gamma symbolisms can simplify an equation. If we make one more step in modernizing Whitehead's formula from above, we get

$$-\partial_\rho \Gamma^\rho_{\mu\nu} - \Gamma^\rho_{\mu\nu}\partial_\rho \log\sqrt{-J} - \Gamma^\rho_{\mu\sigma}\Gamma^\sigma_{\nu\rho} - \frac{\partial^2 \log\sqrt{-J}}{\partial u_\mu \partial u_\nu} = 0$$

Whitehead's formula (19), emphasizing summation convention

This version makes the operation and elegance of both the gamma symbolism and the previously mentioned summation convention much more obvious, because it fully exploits the distributions of indices between the gamma symbols and the partial differentials. If we look at just the first term, the lower index "ρ" in the ∂_ρ corresponds to the upper index rho in the $\Gamma^\rho_{\mu\nu}$ and tells us that this is actually a matrix of elements summed across the rho[th] complex. It is also worth recalling that, while neither the gamma nor the partial differential is by itself a true tensor, the combined formula above is. The reader should compare the above with the corrected version of Whitehead's formula (19) on page 158, to gain some appreciation for just how efficient these various representational choices really are. Because our next step is to learn to appreciate how dangerous they can be for the unwary.

II. Whitehead's Theor(ies) of Gravity:

There are a variety of paths one might take in attempting to gain an insight into the inner workings of Whitehead's theory of gravity. The path that has most often been followed has been one of translating Whitehead's theory into a mathematical format that enjoys a greater level of familiarity amongst physicists and then to explore its consequences from there. This is the method used by Eddington, for instance, when he demonstrated the equivalence of Whitehead's theory with that of special relativity by embedding the former in the "Schwartzchild metric," a standard system of spherical coordinates in relativity theory (Eddington, 192). Synge continued and extended this process, but at a cost. He abandoned

Whitehead's own technique of separating of the metrical relations into two tensors, a "J" term for the gravitational relations and a "G" term for the purely geometrical ones,[111] preferring instead to develop Whitehead's mathematics along the lines of more standard usages (Synge 1952, 310). Following the work of Eddington and Synge, A. Schild generalized Whitehead's theory into an entire class of gravitational theories (Schild, 203). Still another physicist, C.B. Rayner, similarly based his work on Eddington and Synge (Rayner, 509 ff). Again, all of these authors resort to the standard "$g_{\mu\nu}$" symbolism rather than Whitehead's choice of splitting the metric relations between his "J" and "G" tensors, and then building out the distinction between the gravitational and geometrical metrics from there.

Mathematically, at least, there is nothing wrong with such a choice. As long as one is consistent in the use of a particular formalism, then the specific formalism is – *formally* – a matter of no great concern. But we have already seen that there is more to mathematics than just formalism. Intuition and aesthetics play an essential role in the conduct of mathematical inquiry. And both of these, as we have seen, are educated by a variety of factors. Indeed, while it is certainly the case that the standardization of symbolism today makes it impossible to look at "$g_{\mu\nu}$" and not immediately interpret it along the lines of orthodox GR, one may legitimately wonder if it was possible for a mathematician or physicist in even Whitehead's day to see it otherwise. Einstein's reputation within the scientific community was already well established by 1909, and the appearance in 1916 of his paper on general relativity was a powerful influence in determining which usages were going to be the standards. Whitehead was quite aware of Einstein's work, and how Einstein employed his symbols.[112] But Whitehead also repeatedly emphasized that his interpretation of the mathematics in his theory was entirely different from that found in GR. So we should perhaps be open to the possibility that there was more behind Whitehead's choice of mathematical symbolism than merely habit and the legendary, if somewhat apocryphal, English stubbornness around adopting continental mathematical conventions.

This is the elephant in the room that no one quite knows how to talk about. From a purely psychological point of view, it seems almost unthinkable that our habits of thought and orthodox use of symbols could

111. So far, we have only encountered the "J" tensor. It is important and suggestive that "Einstein's Law" can be fully expressed by using only the gravitational half of Whitehead's machinery. We will deal more fully with both in a moment.

112. For example, Whitehead specifically refers his dJ^2 to Einstein's use of ds^2 (R 10).

ever *not* prejudice our approach to a given topic. But despite the fact that some work on this subject has appeared respecting the more empirical branches of inquiry, even twenty years after Davis and Hersh published their groundbreaking volume, relatively little has been done regarding the more "sociological" aspects of research in the formal sciences. Some illustrative examples come to mind, William Hamilton's struggle to make quaternions multiply properly being an especially famous one. Once one knows the trick behind manipulating quaternions, the whole thing seems so straight forwardly obvious that it is hard to imagine how such things could ever have been so mysterious (Kline, vol. II, 776 – 782). But Hamilton did *not* already know how to make these symbols work, and it took him fifteen years to finally surrender the habits of thought he had internalized regarding the nature of multiplication before he could come up with a workable system for quaternions. So we will examine Whitehead's mathematics with an eye toward how they might represent a deliberate distancing of his theory from Einstein's. This will be done in two steps: we will first get an overview of the third part of R, where Whitehead lays out the "fundamental" ideas of tensor analysis, and thus sets out for us what he himself considers most important. Our second step will be a closer examination of Whitehead's formula (19), in terms of how Whitehead himself saw this as possibly generalizing into alternative theories of gravity from both his and Einstein's.

Now, in a subject as rigorously developed – and, as a consequence, as narrowly constrained – as differential geometry, there are only so many different approaches one can take on the topic while still providing anything like an adequate exposition. Some material absolutely must be covered in order for any part of the later developments to even make sense. Consequently, Whitehead begins his treatment of the subject very much where one must begin: with coordinate systems. This presentation is not offered from a purely formal point of view, however. Even though this part of Whitehead's text is called the "Elementary Theory of Tensors," (R 137) the text continues to be dominated by the service it can provide to explicating the *physical* part of his theory. Hence, Whitehead begins by stipulating that coordinate systems with three spatial and one temporal coordinate will be called "pure" (R 139). There is nothing unusual about this focus on systems of coordinates which are not absolutely general, particularly given that the text in which this exposition is appearing is a physical-philosophical one,[113] although Whitehead's commitment to a

[113]. Einstein begins in essentially the same way in his *precis* of tensor analysis in the 1916 essay (Einstein, *et al*, 118).

system of separated spatial and temporal coordinates does stand out. What is even more striking is that Whitehead then goes on to note yet a further potential limitation of perspective which does not as often receive this kind of immediate consideration. Having noted that various scalar quantities of physical importance emerge which are identical between coordinate frames (and taking particular pains to distinguish between such quantities and the *invariant formulae* used to express those quantities), Whitehead also points to certain types of formulae which are not invariant in general, but only within the limited scheme of a mathematical group of transformations (R 140 – 1).

This is a rather interesting move to make. Many texts on relativity and differential geometry give little or no explicit attention to group theoretical structures, despite the fact that the Lorentz *group* is in many respects the cornerstone of relativistic coordinate transformations in both the special and the general theories. At the same time, Whitehead himself makes no specific mention of the Lorentz transformation. In fact, Whitehead's interest in group theory in this context can be traced to one explicit source, and another that, while only implicitly part of Whitehead's thought in R, was certainly a major part of Whitehead's early mathematical education and a matter with which he was substantially familiar.

The explicit element here is Whitehead's reliance upon "Galilean" tensors for the development of his theory. A Galilean tensor is one in which the coordinates are referred to a rectilinear Cartesian system of three spatial dimensions and one temporal. Hence, while a four-dimensional Cartesian system is a perfectly general Euclidean space, a Galilean one is "pure" in the sense Whitehead stipulated above. The Galilean group, then, will be that system of symmetric transformations in which not only the Cartesian structure, but the distinction between space and time is maintained. Whitehead will devote an entire chapter to Galilean tensors and their transformations (R, chapter XXII). The Galilean tensor in Whitehead's theory is the element of pure geometry. It is here that we find the uniform background structure which makes metrical relations knowable to finite minds such as our own. This is his "G" tensor, and we will presently see an example of just how thoroughly entangled this Galilean tensor and the gravitational tensor "J" can get.

The implicit element here is Whitehead's demonstrable familiarity with the topic of spaces of constant curvature. It is to be recalled that Whitehead states on the opening page of R that his interest in Euclidean geometry – and hence, his focus in later pages on Galilean tensors and groups – was a matter of formal convenience only. "I should be very willing to believe that each permanent space is either uniformly elliptic or uniformly hyperbolic, if any observations are more simply explained by such a hypothesis" (R v), a statement we have previously noted, and regarding which it is now worth reminding ourselves. The "uniformly X" spaces Whitehead mentions are the spaces of constant curvature alluded to above. As Moritz Epple has shown, the topic of spaces of constant curvature was an area of vigorous discussion amongst both English and German mathematical circles during the last quarter of the 19th century (Epple 2002, 937 ff). Whitehead himself devoted entire chapters to the subject in his *Universal Algebra*, and referred extensively to the then principle investigators into the subject, the Englishman William Kingdom Clifford, and the German mathematicians Felix Klein and Wilhelm Killing (Whitehead 1898, e.g. book VI, ch.'s 2 – 5). And, to complete the circle, mathematical groups are precisely those functional structures which determine and formalize the "constants" and "uniformities" of curvature in abstract spaces (see, for instance, Helgason 1964, or Wolf 1967). Hence, what Whitehead is showing us in these early pages of his primarily mathematical treatment of tensors is that his focus remains squarely on the pragmatic issues of both applicability *and* uniformity. This latter, of course, is the distinctively Whiteheadian move.

At the opposite end of the spectrum, a move that Whitehead does not make, but which is otherwise universal amongst texts on tensors and differential geometry, is the development of the concept of a "geodesic." One can make such excuses as one may care with regard to the limitations of space confronting Whitehead in the final sections of R, but from the orthodox perspective of either mathematics or physics, failure to treat geodesics as a central topic of concern is nothing less than astonishing. However, it must be emphasized that this astonishment is entirely predicated upon the orthodoxy of one's perspective. Whitehead's perspective is not orthodox, and this apparent lacuna is an expression of his previously stated rejection of the geometric metaphors that were, and continue to be, the dominating influence in the development of tensor analysis. As with his early but explicit mention of groups of transformations, Whitehead does not take the next step and specifically link this choice of formal development with his previously articulated philosophical position. However, just as the previous move demands attention for its notable introduction at such an early point in the presentation, so the topic of geodesics calls attention to itself by its very prominent absence. It stretches credulity to imagine that either gesture was casual or accidental on Whitehead's part, not only because of the contrast created with standard approaches to the topic even in his own time, but because of the direct relevance such moves have to Whitehead's philosophical position.

Much of the rest of the text again follows more traditional lines. Whitehead covers the basics of forming tensors – composition, multiplication, addition, etc. Pure tensors of all upper indices (contravariant tensors) or all lower (covariant tensors) are discussed, as well as the basic rules of tensors of mixed type. Christoffel symbols are introduced using, as previously mentioned, the older style bracket method. Covariant differentiation is also discussed, although here Whitehead's exposition is rather boxed in by his philosophical presuppositions. Most texts motivate this part of their discussion by appealing to the very geometric intuitions which Whitehead has precluded from his own theoretical development. So, while it is generally thought to be of the very nature of the covariant differential that it is a tool for comparing the vectorial relations between an initial point and another which is only slightly displaced from the first, Whitehead leaves the nature of this displacement substantially up in the air. Parallel transport receives no mention, which ties in directly with Whitehead's silence on the subject of geodesics – the two topics are essentially flip sides of the same coin. On the other hand, since Whitehead's treatment here is, once again, conducted entirely within the context of the specific physical theory he is developing, rather than a treatise on tensor analysis from a maximally general point of view, this gap can also be seen as representative of his prior commitment to a Cartesian–Galilean system. Within such a framework, laws of parallels are simply those of Euclid, and the rules of parallels are simply those of ordinary geometry. Covariant differentiation, in this context, does not link up with the geodesic structure of space as it does in GR, but rather with the empirical contingencies of the physical field.

In the last few pages, Whitehead introduces what he calls the Riemann–Christoffel tensor, what we have been calling the Riemann curvature tensor. By itself, this tensor is not of much use to Whitehead, since space in his system is flat and the Riemann curvature is thus equal to zero. But, as pointed out above, if one contracts an upper and lower index on the Riemann, and sets this equal to zero, one has "Einstein's law" for the gravitational field. Using a different letter for his operational tensor ("H" rather than "J"), Whitehead rewrites his earlier formula (19) as (58) (R 185), rearranging the terms slightly, and this time correctly associating the summations. The minor change in lettering employed makes it possible to return to his preferred "J" for his gravitational tensor, while this time taking care to not prejudice the issue with the contracted Riemann. He is then able to set out his gravitational formula, in a more general form than he had earlier developed, and show how by an appropriate choice of changes, the general equation at the end of R reduces to the particular application of his theory.

Whitehead's decision to close his survey of tensors here in many respects complements the closure of part I of R. There too, the gravitational law is presented as both the final point to be observed, and the initial point for further development of a physical theory of space and time. This, indeed, is where most treatments of his physical theory begin their developments. And if it were our purpose here to achieve a substantive grasp of the technical details of Whitehead's theory, then this would also be our next move. However, this is not our purpose. Our task involves only gaining an intuitive insight into the ways in which the formal mechanics of Whitehead's theory express his philosophical commitments. For this purpose, it turns out that there is a better choice of formulae for us to examine. At the end of both parts I and III of R – the philosophical section and the more purely mathematical one – Whitehead took one additional step in his development, and mentioned the possibility for alternative theories of gravity from both his and Einstein's. This is where we will turn our attention next. While this material is much more speculative on Whitehead's part, it is also more self-contained. As such, it will serve our needs well.

This is all but entirely virgin territory. As nearly as this author can determine, no one has ever examined this part of Whitehead's theory at any level of detail. Robert Palter notes these ideas of Whitehead's, but makes no effort to examine them (Palter, 208). The only scientists to make an effort to generalize Whitehead's theory were Temple, Rayner and Schild. Of the three, Temple's work was much more directly related to Whitehead's, being presented within a year of the publication of R and benefiting from Whitehead's presence at the reading. But Temple's approach remained focused on Whitehead's standard gravitational formulae (Temple, 177 and 184), dealing in fairly straight-forward modifications of this by incorporating basic ideas of spaces of constant curvature.[114] Schild's research presents results which are considerably more general than Temple's while retaining a keener focus on the empirical adequacy of the class of theories developed. However, Schild's work deals exclusively with classes of flat spaces only. Moreover, it is not immediately apparent whether or not Schild had any direct acquaintance with Whitehead's work,

[114]. Compare (Temple, 179) and, e.g., (Lawden, 123, exercise 33).

despite the mention R receives in Schild's references. Schild was a close and frequent collaborator with J.L. Synge, and everything in Schild's paper appears very much to be based exclusively on Synge's earlier work on Whitehead's theory. Certainly it is the case that Schild makes no explicit reference Whitehead's own speculative comments regarding generalizations of his approach, nor does Schild's discussion exhibit any implicit awareness of such.[115] Rayner's work is, again, a direct development of the theory as presented by Synge, to include non-static spherically symmetrical gravitational phenomena. As a result, while all of these authors offer important insights and generalizations of Whitehead's theory, their approaches do not provide a sufficiently clear exposition of the essential differences between Whitehead's theory and GR. For this reason, we will turn back to Whitehead's own development.

Previously, we have mentioned how Whitehead's theory is built around *two* principle metric tensors, rather than the single $g_{\mu\nu}$ found in GR. Whitehead's two tensors are his “J” for the gravitational field and his “G” for the relational background that makes metrical structures logically intelligible. However, to genuinely appreciate the extent to which these two tensors meld into a single theory, it is worth looking at how Whitehead saw them generalizing beyond his own specific theory. This generalization was his formula (20)[116] which is:

$$\sum_{\rho}\frac{\partial K^{\rho}_{\mu\nu}}{\partial u_{\rho}}+\sum_{\rho}K^{\rho}_{\mu\nu}\frac{\partial}{\partial u_{\rho}}\log\{-G\}^{\frac{1}{2}}-\sum_{\rho}\sum_{\sigma}K^{\sigma}_{\rho\nu}G\{\mu\sigma,\rho\}^{(u)}+\sum_{\rho}\sum_{\sigma}K^{\sigma}_{\rho\mu}G\{\nu\sigma,\rho\}^{(u)}=0$$

Whitehead's formula (20), pg. 86. Summations corrected from original

Note that Whitehead's gravitational tensor J is now gone. In its place is the geometrical tensor G and a new tensor K. Whitehead suggests two possible definitions for the three index tensor “K” which would then be plugged into (20) above to give the alternative theories of gravity. However, before looking at the first of these, there is a natural temptation to want to modernize (20), including rewriting it in terms of the gamma symbolism, thus eliminating the explicit summations and the brackets for the type two

115. Although it is possible that Schild's generalization of Whitehead's theory, which is rather complicated in its own right, might be shown to mathematically include Whitehead's speculations as a kind of limiting case.

116. I have again corrected Whitehead's erroneous summation convention, which he had carried over from (19).

Christoffel symbols. Such an initial rewrite might look like this:

$$\boxed{\partial_{\rho} K^{\rho}_{\mu\nu} + K^{\rho}_{\mu\nu} \partial_{\rho} \log\sqrt{-G} + K^{\sigma}_{\rho\nu}\Gamma^{\rho}_{\mu\sigma} - K^{\sigma}_{\rho\mu}\Gamma^{\rho}_{\nu\sigma} = 0}$$

Whitehead formula (20), naive version

This would appear to make the indexical relations between the differential equations, the Christoffel symbols, and the still to be unpacked K tensor much more explicit, and as such seems to have much to recommend itself.

But such a move would be a significant error. When we previously translated Whitehead's (19) into the modern symbolism, we compressed his "J" tensor into the gamma symbols, in order to maintain consistency in comparison with Einstein's formulations. But the Christoffel symbols in (20) are not operating on the gravitational tensor. Rather, they are acting upon the uniformity tensor "G." The "naive" modernization of (20) we just proposed completely buries the distinction between these two functions within the common gamma symbolism. This is the problem with the modern symbols which was promised in section one. Whitehead's theory is such that, the only legitimate way in which the gamma style of Christoffel symbol can be used in writing out his formulae is if (1) the operator status of the Christoffel symbol is kept firmly in mind, which means (2) the particular tensor being operated on must be *specifically given*. This, in turn, requires that, for consistency of comparison, the exact same rules be observed when dealing with orthodox GR as well.

To be sure, it is probably unlikely that the above error would ever be committed by either an experienced mathematician or physicist. But by permitting ourselves to walk into the mistake, the need to maintain the rigorous difference between the gravitational metric J, which is a matter of physically contingent relations, and the geometric basis of uniformity expressed by the tensor G is made dramatically evident. This is precisely what differentiates Whitehead's theory from that of orthodox GR, both scientifically and philosophically.

Finally, to see how deeply these two metrical structures can be entangled, let us unpack Whitehead's first potential definition for the tensor K in the above. This is given as:

$$\boxed{K^{\lambda}_{\mu\nu} \underset{def}{=} \sum_{\alpha} G^{\lambda\alpha}_{(u)} J[\mu\nu,\alpha]^{(u)} \sum_{\alpha}\sum_{\beta} G^{\lambda\alpha}_{(u)} J^{(u)}_{\alpha\beta} G\{\mu\nu,\beta\}^{(u)}}$$

First possibility for tensor "K"; Whitehead's (20.ii)

Note that we have both a type one and a type two Christoffel symbol involved in the definition of K. Rather than trying to compress this into the gamma style of Christoffel symbol, which would only lose the differences between the tensors being operated on by the Christoffel symbols, let us expand this fully to reveal its differential structure:

$$K^{\lambda}_{\mu\nu} \underset{def}{=} \frac{1}{2}\left[G^{\lambda\alpha}\left(\partial_\nu J_{\alpha\mu} + \partial_\mu J_{\alpha\nu} - \partial_\alpha J_{\mu\nu}\right)\right] - \frac{1}{2}\left[G^{\lambda\alpha} J_{\alpha\beta} G^{\beta\rho}\left(\partial_\nu G_{\rho\mu} + \partial_\mu G_{\rho\nu} - \partial_\rho G_{\mu\nu}\right)\right]$$

Whitehead's (20.ii) rendered explicit

This makes it clear that there are two simultaneous differentiations going on within this version of K, involving both the gravitational and the purely geometric tensors. Yet, in neither case, can these differentiations be simply separated, the one from the other. Bearing in mind that (20.ii) is to be plugged into formula (20) at each of the K positions, rewriting the indices as needed in order to maintain consistency, then the possibilities for multiple layers of interaction between the two tensors becomes manifest.

Once again, this does not appear to be a possibility that has ever been explored within either the physical or the philosophical literature. Perhaps it would ultimately turn out to be a dead end; perhaps it would ultimately prove to be encompassed by some other form of generalization. Our purpose in mentioning this potential development of Whitehead's theory here is not to explore it so much as to use it as a tool to bring into sharp relief the vital distinction in Whitehead's theory – and, indeed, in any Whitehead*ian* theory – between the physical and geometric relations. By following this particular path, we have also seen that, while radically and irreducibly distinct, these two factors might yet be intermingled in multiply layered and complex ways. All of which renders the task of actually evaluating not only Whitehead's particular theory, but Whiteheadian theories in general, an enormously difficult task. Surveying some of the attempts at this task is our project for section three.

III. Evaluating Whitehead's Theory:

While our primary interest in Whitehead's physical theory is as a systematic example of an historically interesting alternative to GR that does not lead into the measurement problem of cosmology, this theory was also presented as a serious scientific proposal. As such, the possibility needs to be considered that the theory, or even any legitimate modification of it, might simply be refuted by the available empirical evidence. Although Whitehead himself never pursued the theory any further than the publication of R, others have and those pursuits have led to various empirically based arguments that we now need to examine. Two arguments in particular call for our attention here, because both present cases that claim to show that Whitehead's specific theory of relativity has been falsified. We will conclude this chapter with a brief examination of the claims made by these two arguments. In order to soften the otherwise polemical nature of my remarks, I will often refer to these as the "redshift" and the "Earth tides" arguments or critiques, respectively.[117]

Of these two arguments, the Earth tides critique is by far the more famous, and is widely accepted amongst the physics community as providing a definitive refutation of Whitehead's theory. However, this proposed refutation is predicated upon a number of assumptions which are substantive abstractions from the known facts, and have been seriously challenged on those grounds. On the other hand, the claims made by the redshift argument, though not as widely known or even published except on the Internet, might well be more telling than those of the Earth tides criticisms. But there are again assumptions within the redshift argument and limitations in its scope that ultimately leave matters far less decisively settled than might first appear to be the case. I will turn first to the redshift critique, and then offer some comments on the Earth tides argument.

Two factors that make the redshift argument of particular value to us here are that the author, while certainly sympathetic to the overall program of process philosophy, is also not afraid to criticize Whitehead's scientific program where he sees that program as presenting empirical inadequacies. But of greater importance is that this criticism, unlike those of Synge or the Earth tides argument, presents Whitehead's theory in Whiteheadian symbolism, rather than translating it into more modern symbolism which potentially masks Whitehead's underlying philosophical intentions. While this does not render the redshift critiques entirely unproblematic, it does at least demonstrate a commitment to deal with Whitehead's theory not only as an abstract mathematical exercise, but as a complete physical and philosophical approach.

Still, there are a few noticeable historical mistakes in the available versions of the argument that seem to suggest that we should approach it with a measure of caution. For example, it refers to the title of Whitehead's *Principle of Relativity* as, "certainly ambiguous and therefore misleading," a characterization which fails to take into account the philosophical uses of

---

[117] The former argument was presented by Yutaka Tanaka on several occasions, and is currently only available on the web, while the Earth tides argument is due to Clifford Will. See (Tanaka 1985) and (Will 1971a, 1993a), respectively.

the term "relativity" at the time that R was published[118] (Tanaka 1985, 2.2[119]). Later on, one reads that, "Eddington first took notice of Whitehead's theory," (Tanaka, 4.1), referring to Eddington's short piece from 1924. However, as we have already seen, it was Temple in 1923 who first contributed to the physical and mathematical literature of Whitehead's theory. These are relatively minor cavils by themselves, but they do speak to a certain lack of care in the development of the argument which we shall encounter in a more substantive form momentarily.

Other than these few minor exceptions, the early portions of the redshift argument give a very satisfactory gloss of Whitehead's theory. In particular, section two of the article outlines Whitehead's philosophical commitments while some of the more important aspects of the mathematical machinery are accurately spelled out in section three. In these early sections, one also finds that the nature of Whitehead's requirement for uniformity and its reasons are correctly identified. Matter and energy are not essential components of the structure of space and time itself, being (on Whitehead's theory), mere "adjectives" and not fundamental aspects of the relational uniformities of events (Tanaka, 2.12; 2.15 – 2.17). Even more impressive is the handling of Whitehead's mathematics. As already mentioned, Whitehead's own formalisms are presented, and section three of the article very competently walks the reader[120] through the basics of Whitehead's formal machinery. The brevity of the articles requires that some of the minutiae we have examined get left out. Many of the details of the history of Whitehead's theory are skipped over, and none of its immediate alternatives are explored nor is Whitehead's formulae (20) mentioned. And while the argument makes it clear that Whitehead's use of Euclidean geometry was an optional part of his theory (Tanaka, 2.18), no mention is made of Temple's 1923 generalizations of Whitehead's theory to non-Euclidean contexts.[121]. But these, again, are minor limitations in management of the technical details

118. Which is to say, "relative" was taken philosophically to mean what we might now call "relational."

119. Since Tanaka's article is from the Internet, page references are of no value. Consequently, citations will be given in the form of "S.P," where "S" refers to the section number and "P" to the paragraph within that section.

120. Or, perhaps one should say, listener, since this paper was originally delivered at conferences at the Center for Process studies on two different occasions, in 1984 and 1985.

121. Tanaka does cite a paper of Temple's from 1924 which does include much of this previous work (Tanaka, 4.1).

of Whitehead's work, especially given the succinct nature of the overall discussion. In addition, it is clear from the text that the author was aware of the purported disconfirmation of Whitehead's theory offered by the Earth tides argument. But it is also made clear that there remain potential issues around this latter work (as we shall see in a moment) which throw some of its central claims into doubt, and the subject is not pursued.

This brings us to the central point of contention in the article, the gravitational redshift. While the article correctly identifies this as an empirical test which Einstein considered to be of central importance in establishing his theory (Tanaka, 1.4; Einstein, 151), we also noted in chapter one that the gravitational redshift is, at least by some accounts, less a test of GR itself as much as it is a test for the equivalence principle. However, in the early days of GR, this was a refined distinction that no one was yet prepared to make. For Einstein, the redshift was a natural and automatic result of the collapsing together of physics and geometry, for the redshift is nothing else than a direct manifestation of the modification of the metric structure of space in the presence of a gravitational field. This, of course, is exactly what leads to the measurement problem of cosmology, and as a consequence no such interpretation was available to Whitehead. Yet Whitehead's theory also involves effects similar to the gravitational redshift, although the value Whitehead determined from his theory came to $^{7}/_{6}{}^{th}$ of the result from Einstein's (R 11, 102 – 103). One might add that this is despite the fact that at the time of Whitehead's proposal there were no tests available that could effectively determine if such results even existed, much less if they fell more in line with Einstein's predictions or Whitehead's.

However, we now do have quite refined tests for the effects predicted by Einstein and Whitehead, and as the redshift argument points out, these tests work out strongly in favor of Einstein's theory (Tanaka, 5.6), coming to within one percent of the predicted value (Will 1993b, 53 – 4). Whitehead's theory of the redshift was built upon an extremely simplified, pre-quantum mechanical model of the molecule, and so there are some refinements possible that can close the gap between his model and Einstein's to within the margin of observational error. But the redshift argument invites us to consider if such ongoing "*ad hoc*" tinkering is even appropriate. It is not just that the data fits Einstein's theory. The data fits Einstein's theory easily and directly. Whitehead's theory, in both its original and any updated forms, requires various secondary hypotheses about physical interactions simply in order to get its initial predictions for the redshift effects, predictions which GR produces effortlessly (Tanaka,

5.7). Furthermore, the gravitational redshift results predicted by GR generalize to *any* physical situation involving light and gravity. On the other hand, such generalization does not seem to be available to Whitehead's theory in its existing formulation (Tanaka, 5.8;[122] Schild 1961, 81 ff).[123] Thus, the argument goes, Whitehead's theory in its original form gives an extremely limited and limiting account of the redshift phenomena which make the theory appear incomplete and unsatisfactory. The conclusion is that even further modifications are needed if one is to capture this readily generalized aspect of GR. We are asked to suppose that such further modifications are unreasonable and inappropriate.

However, it is at this point that the redshift argument shows some significant problems. For one thing, the last stated claim above is predicated upon the acceptance of the "uniform occurrence of the gravitational redshift" (Tanaka, 5.8), and that Einstein's theory is the more thoroughly scientific because it "runs the risk of being refuted by possible varieties of gravitational redshift" (Tanaka, 5.7). But the first point is an assumption, while the second is readily stood on its head – for example, Whitehead's theory runs the risk of falsification in virtue of a too uniform phenomenon of gravitational redshift.[124]

Moreover, Whitehead's theory is not as *ad hoc* in its development of the connection(s) between matter and electromagnetism as the redshift argument (or Schild, in his earlier piece) would seem to suggest. Whitehead's philosophy mandates a different approach to the relations between physical phenomena and the metrical relations of space, which does not admit of the kinds of tidy reductions one finds in GR. This does not make the additional complexities of Whitehead's theory somehow *ad hoc*. These complexities are representative of the philosophical commitments of Whitehead's theory, commitments that make the underlying theory of measurement logically viable. One can as readily counter with a similar accusation against GR for its cavalier assumption of the *ad hoc* possibility of cosmological measurement in the absence of even a possible grasp of the logical relations that would actually make such measurements meaningful.

---

122. Tanaka's citation for Schild is slightly inaccurate, putting the year of the publication at 1963, rather than the correct date of 1961.

123 In this regard, (Schild 1961) argued that the spatial metric within Whitehead's theory was otiose and ad hoc.

124. Bear in mind that, since the redshift is associated with contingent physical relations in Whitehead's theory, there is no reason why it should manifest the uniformities required by geometry to resolve the measurement problem.

The redshift argument is further weakened by the ambiguous manner with which it handles Whitehead's openness to geometries of constant curvature as the basis of the uniformity of space. On the one hand, the argument cites the very passage from R which we have had occasion to quote, wherein Whitehead asserts his willingness to believe space is uniformly curved (rather than Euclidean or flat), "if any observations are more simply explained by such a hypothesis" (Tanaka, 5.4). Yet, on the other hand, when the redshift argument presumes to explain the limitations of Whitehead's theory in dealing with the gravitational redshift – arguments which are cited as coming directly from Schild – the critical claims it presents are explicitly related to the *flat* space used by Whitehead to exemplify his theory (Tanaka, 5.8 ff). The entire concluding portion of the discussion appears to take it for granted that the problem with flat spaces and the assumption of a generalized gravitational redshift is directly contrary to Whitehead's theory at large, despite the earlier awareness that flat spaces, for Whitehead, are not a mandatory feature of that theory. And, again, all of this takes for granted the assumption that the gravitational redshift is uniform throughout all space and time – or, as in Whitehead's theory, all space*s* and time*s*. To what extent these redshift issues generalize to spaces of constant curvature is not explored nor even noted.

Finally, the position of the redshift argument is that its criticisms of Whitehead's specific, applied theory constitute global refutations of Whitehead's larger critique of GR. So even as it takes notice of the metrical issues in Whitehead's work (Tanaka, 2.12 ff) – what we have been calling the measurement problem of cosmology – the argument does not appear to grasp their fundamental significance either for Whitehead's theory or cosmology as a whole. Rather, on the basis of the redshift argument's criticisms, we are asked to conclude that a Whiteheadian should be prepared to abandon the entire program of R, and be content with something like uniform *topological* relations instead of the much stronger metrical uniformities required by Whitehead. "Adopting Whitehead's paradigm," the argument states, "we can require that only the topological structure of space-time should be independent of matter, and thus a priori relative to measurement" (Tanaka, 2.18). Or, again, taking the point of his criticisms as valid, it is asserted that, "we must take the other alternative. i.e. the metrical properties of space-time are indeed affected by the existence of matter, but the topological properties are independent of matter" (Tanaka, 5.15; punctuation in the above is uncorrected from the original piece). It goes on to state that, "Whitehead himself was absorbed in the topology of events as the 'theory of extensive continuum' in his later philosophy of nature" (*ibid*). Yet the argument fails to notice that, even while thus "absorbed," Whitehead continued to explicitly endorse his criticisms of the *metrical* problems in GR (PR, 332 – 3; and part IV, chapter V in general). The reason for the continued emphasis on the metrical properties of space, and not merely their topological characteristics is simple: there is no way to derive metrical properties from pure topology alone. A casual glance at any text on the subject makes this fact apparent. Metrical relations are never *de*duced, they are *intro*duced. No amount of topological uniformity by itself will suffice to produce the needed structures to resolve the measurement problem of cosmology.

The failure of the redshift argument to deal effectively with the larger context of Whitehead's theory is not unique, but it is disappointing. Still, at least we see in it an attempt to deal with Whitehead on Whitehead's own terms. Such a procedure does not seem to be the norm in the physics literature regarding Whitehead's theory. We have seen how J.L. Synge's discussion failed to take Whitehead's ideas on their own terms, but rather translated them into more familiar mathematical formulae. The Earth tides argument, which is based extensively upon Synge's formulations, suffers this same problem. In addition, and unlike the redshift argument, Whitehead's philosophical discussions have been entirely left behind. Still, the Earth tides critique offers another potential source of an empirical refutation of Whitehead's theory. Let us briefly examine this claim.

The purported refutation offered by this argument is described easily enough. It is a consequence of Whitehead's theory that there will be tidal forces on the earth related to the distribution of matter in the galaxy. In principle, these "earth tides" make for a measurable difference between Whitehead's theory and GR. These earth tides – a consequence of the "gravitational anisotropy" due to the separation of the gravitational and the spatial metric – appear because the fixed geometric background of Whitehead's theory does not permit one to redefine the frame of reference at will to the extent that one is able in GR (Will 1971b, 149). This result is not dependent upon whether or not Whitehead's theory is interpreted within its original "flat" spatial structure. It is the fact that the geometry is *uniform* and the gravitational forces in Whitehead's theory are propagated in straight lines, which means that these tidal influences from the rest of the galaxy cannot be defined away by translation into a new frame of reference (*ibid*). With this in mind, the Earth tides argument assumes certain functional relations regarding the structure of the galaxy, and in this form can be used to generate a numerical estimate of what these tidal forces must amount to. Experimental tests for such forces show that, if they exist, they must be within a three percent margin of error of zero. But the standard calculations of the Earth tides argument show that Whitehead's theory predicts effects some 200 times greater than anything measured (Will 1971b, 141 – 3). If these calculations are correct, Whitehead's theory cannot be true.

But there is a problem with these calculations. As Dean Fowler has pointed out, it has been assumed for purposes of simplicity that all the matter in the galaxy is concentrated in a point at the center of the galaxy, some 20,000 light years distant, and all of the tidal effects radiate from that point (Fowler 1974, 288). Such an idealization is common enough in physics, but it is only legitimate so long as the simplification does not significantly effect the comparison between observed and theorized results. Fowler claims that, "with a more realistic model in which the mass is smeared throughout the galaxy, Whitehead's prediction is altered by a factor of 100, greatly diminishing the divergence between his prediction and Will's experimental limit" (*ibid*). Indeed, if Fowler's estimate is correct, and one further takes account of the distributions of matter and energy throughout the universe at large, then Whitehead's theory is even closer to experimentally measured effects.

But here we must note an especially frustrating aspect of Fowler's claim: while it is certainly reasonable that a more realistic model of the galaxy and universe would reduce the number that has been calculated, Fowler himself never provides us with his calculations showing how he came up with his asserted two orders of magnitude improvement over the Earth tides calculations. This is true not only of all of Fowler's published articles, but of his dissertation as well (Fowler 1975), which otherwise deals at some length with the formal aspects of Whitehead's theory.[125] So, in the absence of some calculations of our own, Fowler's assertion must be treated with caution. While it has a measure of plausibility, augmented by Fowler's own competent handling of relativistic mathematics, it remains nothing more than a plausible claim. But it is a claim that we will not pursue any further. Were it the case that a crucial decision between Einstein and Whitehead hung in the balance over this single point, then the matter would take on a much greater importance than it does, in fact, possess. For while it is certainly worth noting the above issues, and the possibilities they present for further scientific inquiry, we must also note that the philosophical issues raised by Whitehead's applied theory of

125. One should also add that, Fowler did not stay in academia after completing his Ph.D. at Claremont Graduate School. As a result, his published materials on this subject are quite limited.

gravity in no way hang upon the truth or falsity of the Earth tides critique.

There are two constellations of factors contributing to this global lack of absolute relevance regarding the ultimate viability of Whitehead's applied, scientific theory. As to the first of these, it was pointed out in chapter three that there are other bimetric theories of gravity besides Whitehead's, some of which are still considered viable. These theories are not only viable scientifically, they are also alternatives which meet Whitehead's criterion for a uniform background geometry that could avoid the measurement problem of cosmology. Now, in that earlier discussion, the case was presented that Whitehead's theory should itself be classed as a bimetric theory rather than a quasi-linear one, as Clifford Will has done (Will 1993a, 138 ff). Let us now reiterate that claim by examining some further points of contact between those theories which are classified as bimetric within the physics literature, and Whitehead's proposal.

The first point to note in this regard is that it is Nathan Rosen who is commonly cited (Piso, *et al* 1994, 1) as the originator of bimetric theories of gravity for his articles, published in tandem in 1940 (Rosen 1940, 147 – 153). Rosen worked on ideas relating to bimetric theories of gravity for over thirty years (e.g., Rosen 1974). In Rosen's own formulations, his theory ran into trouble in its predictions regarding gravitational waves and the comparison of his theory with observations of pulsars, and the failure of his theory to predict black holes (Piso, *et al*, 1 – 2). However, modifications of Rosen's work by Piso, Ionescu-Pallas, and Onofrei brought that theory back into a viable condition (*ibid*), and on this account one can still legitimately talk about Rosen's work as a live option in physics. This is of particular interest to us because of how Rosen's speculations compare with Whitehead's.

Rosen begins by introducing two fundamental metric tensors, a gravitational metric $g_{\mu\nu}$ (which is not the same as GR's $\mathsf{g}_{\mu\nu}$) and an abstract metric $\gamma_{\mu\nu}$ which imposes a Euclidean structure at each point in the space in addition to the Riemann metric of $g_{\mu\nu}$ (Rosen 1940, 147). Rosen notes that certain convenient features arise from such a formulation. Among these is the fact that Rosen's formulation greatly simplified the process of differentiation when compared to GR, where one had to work with linear approximations when making actual calculations (Rosen 1940, 149). On this point, it is particularly interesting to note that Rosen's theory and the modified version developed by Piso, *et al*, is, like Whitehead's, a *linear* theory of gravity (Piso, *et al*, 1). As Whitehead notes regarding his own gravitational formula, because it is linear, calculations within his theory are vastly more tractable than those within the non-linear structures of GR (R

84). Returning to Rosen's argument, having presented a variety of formal advantages to his bimetric approach, Rosen speculates a little more about the physical meaning of the two metrics in the second of his two articles. He states,

> From the standpoint of the general theory of relativity, one must look upon $\gamma_{\mu\nu}$ as a fiction introduced for mathematical convenience. However, the question arises whether it may not be possible to adopt a different point of view, one in which the metric tensor $\gamma_{\mu\nu}$ is given *a real physical significance as describing the geometrical properties of space*, which is therefore taken to be flat, whereas the gravitational tensor $g_{\mu\nu}$ is to be regarded as *describing the gravitational field* (Rosen 1940, 150, my emphasis).

At this point, the sense of *deja vú* should be overwhelming. But matters do not end here.

Rosen also points out that the introduction of the metric tensor $\gamma_{\mu\nu}$ (which bears obvious analogies to Whitehead's G) opens up the possibility of alternative laws of physics (Rosen 1940, 150), again paralleling Whitehead's work. Moreover, Rosen notes that, "the velocity of light will, in general, be different from that in field-free space," thus surrendering the absoluteness of that velocity. This is not a topic we have had much call or opportunity to discuss, but it is the case that Whitehead rejected any absolute character for the speed of light, viewing that speed as at most parasitic upon a theoretically postulated critical velocity "c" (R 36; 76). It is interesting to note that bimetric theories of gravity seem to lend themselves quite well to such a supposition, which in turn is quite useful in the explanation of certain physical phenomena (Moffat 2003, 282 ff).[126] This was a step that Whitehead never explicitly took, but which is clearly an open possibility as far as his scientific and philosophical commitments are concerned.

Finally, besides noting the uniformization of measurements that occurs in his bimetric theory, Rosen offers the following observations:

> In the Einstein general relativity theory gravitation is explained in terms of geometry. In the theory suggested here ... this geometrization of

126. Rosen also mentions some possible advantages to a varying speed of light theory, and a third article discussing this possibility followed the tandem articles on the bimetric theory. However, given the extreme fluidity of experimental results, little is to be gained by pursuing a troubling observation first published in 1933. See (Rosen 1940, 150).

> gravitation has been given up. Perhaps this may be regarded by some as a step backward. It should be noted, however, that the geometrization referred to has never been extended satisfactorily to other branches of physics, so that gravitation is treated differently from other phenomena. It is therefore not unreasonable to wonder whether it may not be better to give up the geometrical approach to gravitation for the sake of obtaining a more uniform treatment for all the various fields of force that are to be found in nature (Rosen 1940, 150).

Rosen hints at an issue we will be looking at in the final chapter, the problem of reconciling relativity with quantum mechanics. But the potential usefulness in this regard has been explicitly mentioned by others. J.W. Moffat notes that his own bimetric theory of gravitaty, "can be the preliminary stage for a quantum gravity theory that does not conflict with some of the basic properties of quantum mechanics," (Moffat, 283).

The second matter which renders the above question of viability regarding Whitehead's applied theory of questionable relevance is the fact that even orthodox cosmology – perhaps especially orthodox cosmology – is facing a crisis. Observations of the last ten to fifteen years in particular are proving increasingly difficult to explain, and it is no longer evident that we have *any* viable theory of cosmology. If Whitehead's applied theory is no longer workable, it is in good company. This, of course, makes Whitehead's philosophical criticisms of GR all the more apposite. Examining a few of these issues is our project for the concluding chapter.

CHAPTER SIX
The Big(ger) Picture

> *But consciousness proceeds to a second order of abstraction whereby finite constituents of the actual thing are abstracted from that thing. This procedure is necessary for finite thought, though it weakens the sense of reality. It is the basis of science. The task of philosophy is to reverse this process and thus to exhibit the fusion of analysis with actuality.*
>
> Alfred North Whitehead, "Mathematics and the Good"

I. Why should a Physicist listen to a Philosopher?

Let us review our argument to this point.

In chapter one, the idea was presented that there is a measurement problem of cosmology which is inherent in the very structure of orthodox general relativity (GR). The logical, as opposed to simply empirical, nature of the problem was discussed. Whitehead's triptych of works on the philosophy of nature was introduced as our opening onto the problems of GR. In particular, attention was drawn to Whitehead's 1922 *Principle of Relativity*, less as a source of a complete alternative theory of gravity as an exemplar of a kind of theory which consciously avoids the measurement problems associated with GR. The idea was presented of engaging a formal theory in much the same way as one would engage any other text written in a foreign language. Perfect understanding is not necessary so much as a willingness to grapple with the unknown, and an openness to the process of familiarizing oneself with alien techniques for representing patterns of relations. Some of the aesthetic, intuitive and analogical aspects of mathematical inquiry were mentioned, as a preliminary to our sojourn (beginning in chapter two) into the formal and historical background to GR.

In chapter two, we examined some of the formal and conceptual structures of GR, including on the one hand such basics as the Lorentz transformation, and such exotica as, on the other hand, the identification of space and gravity with each other within that theory. The former gave us our first contact with issues of symmetry, while the latter opened the door to the core of the measurement problem. However, it was the problems associated with symmetry, and the lack thereof in the older theories of space and time, that led us into the important historical and aesthetic issues that were central contributing factors in the acceptance of GR as the cornerstone to cosmology. For many decades, Whitehead's theory could not be empirically distinguished from that of Einstein's, but it received only the tiniest shred of attention from the scientific community. While it is still not entirely clear what the scientific standing of Whitehead's specialized, physical theory might be, we also learned that there are a variety of other theories of gravity, some of which are not only viable, but are bimetric theories like Whitehead's. As such, these other theories fulfill Whitehead's philosophical requirements, regardless the status of his own scientific theory of gravity.

In chapter three we examined some of the mathematical ideas central not only to GR, but of vital importance in any workable physical theory of space and gravity. We saw a little of how tensors operate, and how they are interpreted geometrically. Because of the essential role played in GR by the metric tensor, this geometrical interpretation of tensors is what makes for the intimate connection between gravity and space. We also saw how the contemporary PPN formalism developed by Clifford Will and others establishes the context in which all metric theories of gravity are now compared. But a metrical theory does not explain how those metrical relations are discovered or known, it only provides a formal system in which numbers are generated that can be compared with empirical results. It is still the case that a theory of measurement is needed in order for those metrical structures to stand in a coherent relation to the possibilities of knowledge available to finite minds.

Such a set of systematic logical relations as form a theory of measurement stand as the "natural" correlate to a theory of nature itself. This was the subject of chapter four, where we saw how Whitehead's theory of nature, unlike the "Parmenidean block" universe of GR, provided a theory of nature that was at once able to take seriously the structures of human experience while also providing the outline of a theory of abstraction that could account for the effectiveness of mathematical models in science. By the employment of the method of extensive abstraction, the *idealized* structures of mathematical physics emerge from the *realized* structures of extensive nature, as these are found in experience. We sketched out some of the salient features of Whitehead's theory of extension, and showed how the congruence relations needed to anchor the logical theory of measurement could be derived from the more primitive relations of extension and cogredience. Finally, it was pointed out in the addendum to that chapter that, while Whitehead continued to evolve the details underlying his theory of extension, the fundamental nature of that theory and its role in his critiques of the Einsteinian view of nature remained constant to the end of his career. Moreover, that theory of extension is the basis for ongoing research into such exotic formal topics as "mereotopology," and the comparatively applied domain of spatial reasoning within the context of artificial intelligence.

Finally, in chapter five, we examined some of the formal details of Whitehead's physical theory. We were confronted by a variety of textual issues, including arcane symbolism, errors, and stylistic differences amongst texts, that complicated the reading of his theory. However, as we worked through some of these problems, two points in particular became quite evident: Whitehead's theory is, on the one hand, thoroughly bimetric in nature while, on the other hand, the nested layers of interactions that Whitehead saw as possible avenues of development show that we cannot presuppose a simplistic system of relations between these two metrics as in any way essential to a Whiteheadian style of gravitational theory. However, even at the most superficial level of Whitehead's specific physical proposal, we found that the criticisms and purported refutations continue to suffer from serious weaknesses. Moreover, there are other bimetric theories, developed independently from Whitehead's yet motivated by many of the same considerations, which in various formulations continue to present themselves as viable alternatives to GR. The unfortunate separation of these theories from Whitehead's – the latter being somewhat misleadingly classified as "quasi-linear" in contrast to the bimetric classification which has been reserved for the former group of theories – has weakened the development of both branches of inquiry. Those theories classed as bimetric are left dangling in an interpretive limbo for want of an adequate philosophical basis, a basis which stands ready for them in Whitehead's theory, but which never gets brought into consideration. On the other hand, the range of empirical possibilities for Whitehead's theory are narrowed to the most limited interpretations available, due to the failure to see it appropriately in a robustly bimetric context.

Still, for all of this, a person might well be willing to concede the correctness of the preceding argument, and yet respond to all of the above with a considered, So what? After all, the objections we have raised to GR amount to conceptual and epistemological haggles. What if it just is the case that the world is as Einstein said it is? What if general relativity just is the true model of nature? Why, in particular, for all of the above, should physicists be concerned over our philosophical quibbles, when they evidently have a working theory?

There are two responses to be presented here. The first, in keeping with the fact that this is a philosophical investigation, is a reiteration of the fact that our concern here is not merely with the contingent matters of what we *do* say about nature, or the evaluative issues about what we *ought* to say about nature. Rather, our concern is with the logical issues about what we *can* say about nature. A physical theory which undercuts the logical basis of measurement *cannot* provide measurement based knowledge. In its basic form,[127] it is not possible to know that the universe is as general relativity says it is, because general relativity makes such knowledge impossible. Indeed, one does not have to pull the curtain back very far on any defense of GR (and criticism of our argument here) to see a "spectator theory of knowledge" lurking in the wings. The style of the question posed presumes that there is a "way the world is" that is independent of the means, methods, and possibilities of our knowing what that way is. It presumes, in other words, that there is a "God's eye view" on reality from which all questions about the "nature of Nature" can be answered in a fashion that bears no intrinsic connection to the logical and epistemological relations with which that nature comes to be known. But to make such an assumption does not avoid the need for such connections. Rather, it merely ignores them and hopes a miracle will come along to connect them for us.

Miracles might happen, but they scarcely constitute a reliable methodology. Ultimately, the only way we can know that the universe is any "way" at all, is if our methods of knowing are themselves logically coherent. One element of that necessary coherence is that those methods must at least admit of the possibility of success, however poorly they and we might fare in actual practice. If the universe is "really" as it is represented in GR, then our theory of measurement is built upon a vicious logical circle and cosmology is once again left relying upon a miracle in order to give any meaning to its practice. This is because the metric tensor of GR, which defines the measurable properties of each point in space, is itself varying in response to all of the material and gravitational influences effecting that point. So in order to know the metrical properties at that point, one must first know all of the influences contributing to the metrical tensor. But in order to know all of those influences, one must first be able to make reliable measurements, measurements which presuppose a prior grasp of the relevant metrical structures of the the contingent physical relations of space. In other words, before one can know anything, one must first know everything. On the other hand, if cosmology is a meaningful pursuit – and what useful purpose could be served by assuming otherwise? – then our cosmology must be a theory that can be made sense of from

127. This is a very tricky caveat which must be dealt with carefully. We will spend the final two thirds of this chapter exploring a few paths through this minefield.

those very logical and epistemological considerations upon which it is built. There must be humanly accessible relations which make measurement possible.

But despite these logical and philosophical arguments, it might still be thought that we have only chased ourselves around in a circle. Working scientists simply are not going to spend any effort fretting over a philosophical cavil when they have a functioning physical application in hand. As Whitehead himself pointed out, "when all has been said respecting the importance of philosophy for the discovery of scientific truth, the narrow-gauged pragmatic test (of a working theory in hand) will remain the final arbiter," (R 6 – 7). But it is essential to recall here that the connection between cosmology and a geometrized theory of gravity is not predicated in any essential respect upon that geometrical structure. As we have seen above other theories, which might include Whitehead's, are perfectly viable alternatives to GR and manage this by treating gravity in a purely algebraic manner. Again, we have also seen that Whitehead was not alone in questioning the use of geometric metaphors for gravity: Nathan Rosen was willing to speculate in print over whether or not such ideas were positively obstructing the development of a unified physics. The geometrization of gravity is not a requirement of a functioning cosmology, and that it has so thoroughly entrenched itself at the heart of macro-physics is hardly more than an accident of history, an artifact of how Einstein chose to express his ideas. We have shown that other expressions are not only possible, but they demonstrably exist.

This artifact of expression is itself an example of a confusion Whitehead discussed, between a general idea and its specific applications. The thoroughgoing relational connectedness of space and time is a fundamental general insight into the workings of nature, and it is an unqualified testament to Einstein's genius that he so forcefully expressed this connection, and developed it in his scientific arguments. But the brilliant *general* insight in the relatedness of nature has become hopelessly entangled with the *specific* form in which Einstein chose to present that scientific work. Speaking of the situation in his own time, Whitehead could none the less be addressing physics in our own in that they, "like the Israelites when they fled from Egypt, ... borrowed their valuables – and in this case, the valuables were certain root-presuppositions respecting space, time, matter, predicate and subject, and logic in general" (R 6).

So despite the fact that there are perfectly viable alternatives to GR which entirely side-step the catastrophic philosophical difficulties of that latter theory, it requires something far more effective than mere philosophical argument to engage in a serious reconsideration of our basic assumptions regarding nature, space and gravity. It is significant, then, that there are essential respects in which it also appears that our orthodox cosmological theories which are built around GR, are failing the scientist's narrow-gauged test of "working" as well as the broader gauged standard of conceptual coherence. Some examples will help make this point more evident. However, our first example is more of a *success* story for cosmology, and as such, an important cautionary tale for us about the dangers of leaping to conclusions on the basis of empirical evidence in its early stages of development.

- *The Hubble Constant*: One of the primary missions of the Hubble Space Telescope (HST) was to provide a refined estimate of the Hubble constant ("H_0"), a numerical value essential in the calculation of intergalactic distances and age estimates of the universe based upon observed redshift spectra of various objects in space. Known as the "Extragalactic Distance Scale Key Project," the intention of the project was to improve on the accuracy of the then available ground based estimates of the Hubble constant, which had margins of error in the 20% - 30% range, to a more respectable 10%. However, the initial results from the Key Project gave a value to H_0 of 80 ±17[128] which is quite high. In conjunction with zero or low values for the cosmological constant, this would place the age of the universe at 12 billion years (12 Gyr) or less, which is younger than the calculated age of globular clusters in this galaxy (Kennicutt, *et al* 1995, 1489 – 91). Having stars in the galaxy that were older than the universe would clearly be a problem for cosmological theory. However, continued HST observations and refined methods of combining results from multiple sources led to a final H_0 estimate of 72 ±8. This, in conjunction with slightly higher values for the cosmological constant (for which there is independent evidence) put the age of the universe closer to 13 Gyr and the ages of the globular clusters at 12.5 Gyr. In addition, there is great confidence that the intended range of error of 10% was achieved (Freedman, *et al* 2001, 69 – 70).

Hence, a success story for cosmology. But success is never final in science. Because the refinement of the Hubble constant ideally makes our distance measurements of extragalactic objects more robust, the following examples are for their part made all the more troubling:

---

128. The actual units which these numbers represent are not of importance to us, only the comparison of the numbers from the earlier to the later studies.

- *QSO in Galaxy NGC 7319*: As mentioned above, the Hubble constant in conjunction with the observed redshift in the electromagnetic spectra from extragalactic objects is used to calculate the distances of those objects from us. The higher the redshift, the greater the distance. However, when investigating a powerful x-ray source (a "ULX") detected by the Chandra x-ray observatory[129] in the relatively low redshift galaxy NGC 7319, a high redshift "Quasi Stellar Object" (QSO, or "quasar") was discovered that was evidently interacting directly with the gaseous material of the galaxy (Gallianni, *et al* 2005, 88 ff., 93). By standard calculations based upon its redshift, the QSO ought to be at cosmological distances, when in fact it appears to be embedded in a relatively nearby galaxy. If this observation holds up to further scrutiny, then the entire framework of estimating extragalactic distances could be called into question.

- *Abell Cluster Bulk Flow*: Studies indicate that there is a massive movement of galaxies, including our own, in a common direction and speed, but there is no observable cause for this movement. The volume of space involved in this movement is astonishing, reaching out some 150 h^{-1} Mpc, or about 600,000,000 light years in all directions. In contrast, the so-called "Great Attractor" bulk flow, which received a fair amount of attention in the popular press in the late 1980's and early 1990's, while massive, amounts to a considerably smaller volume of space on the order of 60 h^{-1} Mpc.[130] Unlike the case of the Great Attractor, this constitutes a shocking asymmetry in the movements of galaxies within a staggering expanse of space, and we have no models of cosmology that can account for this (Strauss, *et al* 1995, 507 ff). Again it is the case that these observations are predicated upon an acceptably accurate estimate of the Hubble constant, and an adequate understanding of the nature and meaning of observed extragalactic redshifts.

Not every problem in cosmology is discovered in the context of the unimaginably distant. One of the most profound issues to emerge recently is discernible in the comparative backyard of our own solar system:

- *The Pioneer Anomaly*: Launched in March of 1972 and April of 1973,

---

[129] An orbiting x-ray telescope that began operations in 1995.

[130] For a popular discussion of this latter by one of the main scientists involved in its discovery, see (Dressler 1994).

the Pioneer 10 and 11 space probes were the first terrestrial science packages to reach the outer planets and, eventually, to leave the solar system. Unlike the later Voyager spacecraft with their triaxial guidance systems, the Pioneers were spin stabilized, much like a gyroscope. This design choice made the Pioneers extremely sensitive to external gravitational forces, whose influences on the twin probes could be calculated and measured with enormous accuracy. Based upon a continuous stream of orbital analysis data which ran from 1980 until 1998, and semi-regular contact with the Pioneer 10 spacecraft up until 2002[131], an anomalous effect in the motion of both spacecraft was discovered. The Pioneers have been consistently showing the slowing effects of an acceleration counter to their direction of motion and toward the sun, on the order of 8×10^{-8} cm/s^2 (Anderson, *et al* 2002, 1 ff). While this might not seem like a very large number to a lay person, it is in fact a vastly greater quantity than any force or interaction for which our theories can give any account. The authors point out, for instance, this value amounts to "fit levels (that) are as much as 50 times above the fundamental noise limit of the data. (Anderson, *et al*, 44). While the authors refuse to argue for any need to engage in major theory revision, by their own discussion it is clear that their caution in this respect almost pushes the bounds of good taste. That an internal systematic problem could have caused this exact same acceleration toward the sun on both spacecraft simply defies comprehension, while all known external forces have been precisely calculated and eliminated.

The above examples are all essentially bound up with technological developments which have carried our powers of observation beyond the surface of the Earth. Our final example is more an extended reminder of a fact we have mentioned previously, and which is again more of a conceptual issue than an empirical or observational one. However, the conceptual matter involved is that of the physicists' own domain, and as such, carries more than "merely" philosophical weight.

- *The Irreconcilability of Micro and Macro Physics*: To say that the theories of quantum mechanics and general relativity do not work together is a rather extravagant understatement. Put bluntly, in their present formulations they cannot both be true. This is not a new

---

131. Contact was lost with Pioneer 11 in 1990, while the last weak signal was detected from Pioneer 10 early in 2003. The authorized history of these spacecraft can be found in (Wolverton 2004).

problem. As we saw in the previous chapter, Nathan Rosen speculated upon this matter when he wondered if the geometrization of gravity in GR, a move which at least *was* unique in physics, might not be part of the problem. With the development of modern "string" theories this geometric move may be spreading into other realms of speculative physics as well. However, the structural features of various different string theories and the techniques for translating between these is so utterly contrary to anything like our ordinary intuitions of space and time[132] that one might legitimately wonder in what sense, beyond that of the most abstractively over-extended analogy imaginable, these theories should be characterized as geometrical at all. Furthermore, it is scarcely possible to overstate just how extraordinarily speculative these developments are at this stage. For all of their mathematical cleverness, string theories have yet to produce even the hint or outline of a functionally practical *empirical* test. Until such time as an empirical program of test is at least conceivable, string theories will amount to little more than comforting – albeit, wildly complicated – forms of entertainment.[133]

On the surface, at least, it might seem less than charitable that critics such as Synge and Tanaka would challenge Whitehead's *Principle of Relativity* on the grounds that it was "merely" concerned with, or at least too caught up in, conceptual or philosophical issues, while such an extraordinarily abstract body of work such as string theories are excitedly passed along as good physics (admittedly, by a different group of thinkers). However, we should recognize that – again, on the surface – string theories, by virtue of the fact that they are rigorously formulated mathematical systems, admit of the possibility (albeit, as yet unimagined) that a meaningful empirical test might be devised some time in the future. Indeed, none of the examples above can be supposed to definitively establish the doom of orthodox cosmology. And while each of those examples illustrates the fact that the standard theories do not exercise

---

132. As Brian Greene points out, while different string theories will give more or less identical *analytical* results, the translation of one string theory into another requires moving between completely different underlying "geometrical" structures (Greene, 474 ff). The suspicions toward these "geometrical metaphors" voiced in the above are my own, and not Greene's.

133. For a serviceable lay introduction to string theories, see (Greene, ch. 12 & 13). For a more technical presentation, see (Zwiebach 2004). Again, while the need for empirical test is indisputable, the above criticisms are my own.

complete command of the field, it is also probably true that it is too early in the process to be manning the barricades and singing songs of revolution.[134]

But as problems continue to accumulate – and in any active science, they cannot help but do so – the need to consider extensive revision of the current theoretical framework of cosmology will itself become more acute. When and as this occurs, a re-evaluation of physics in light of the measurement problem of cosmology may well prove to be of considerable importance, one of those sources of data Whitehead himself mentions, which science must turn to when engaged in fundamental critiques of its underlying hypotheses. So let us turn back to this issue, and see now where we stand.

II. Confronting the Measurement Problem of Cosmology:

The only way it could be shown that Whitehead's criticisms of GR's measurement issues were simply invalid is if it could be shown that the logical congruence relations which give meaning to measurement were somehow accessible to us, despite the fact that all geometric relations within that theory are themselves rendered dependent upon the inaccessibly contingent distributions of matter and energy throughout the universe. In other words, it must somehow be possible to beat the problem of having to know everything before we can know anything, while retaining the theory which forces this problem upon us. Whitehead saw the resolution of this problem in the uniformity of nature. Thus, for instance, he argued that, "uniformity of change is directly perceived" (CN 137). In addition, recollecting some terminology from chapter four, there is an, "essential uniformity of the momentary spaces of the various time-systems, and thence ... uniformity of the timeless spaces of which there is one to each time system" (CN 194). Physicists themselves are aware of the general problem posed by the need to know everything before one can know anything. However, one does not see them discuss this issue concretely in the context of GR and the contingency of geometric relations it imposes (see, for example, Greene, 327 ff). In particular, nothing constructive appears to be offered in the physics literature that can, at last, meet Whitehead's arguments against GR directly. Without such a constructively concrete discussion, we are obliged to take Whitehead's criticisms as roughly valid,[135] and are left wondering what we have to fall back on

134. Appropriate nods and apologies here to the ghosts of both Victor Hugo and Thomas Kuhn.

135. The reason for the qualification will become apparent in case number three, below.

beyond mere wishful thinking in order to make cosmology work. Given, as we have seen, the completeness with which GR identifies geometry and gravitational theory, it seems highly unlikely that any such trump to Whitehead's criticism will be forthcoming as such. So accepting Whitehead's criticisms as essentially sound, we are evidently left with three cases, broadly speaking, to consider.

Case number one is that GR is incorrect, and that the "true," or at least scientifically better, theory of gravity is one that is generically Whiteheadian in character. Once again, what we mean here by a theory being generically Whiteheadian is that, in contradistinction to the standard but inadequate habits of classification found in the physics literature, the theory is bimetric in nature. Thus, we must include in our considerations modified versions of Rosen's theory, proposals made by Rastall, and other current ideas such as have been cited in previous chapters, as well as Whitehead's explicit theory and the formal variations on it proposed by Temple, Raynor and Schild. This represents, of course, a very large class of theories, and for this to provide an acceptable alternative to GR, something like a single working model would have to be chosen as "The" theory of gravity at the heart of cosmology.

Presumably such a choice could be made if the empirical pressure to reject GR came to be perceived as meriting such a move. And certainly any bimetric theory which separated the uniform structures of geometry from the contingent relations of physics (as all presently viable bimetric theories do) would serve as an acceptable theory of gravity from a Whiteheadian perspective. But the choice itself would have to be motivated by more than just the theories ability to fulfill Whitehead's criterion for a coherent theory of measurement. There would also have to be a respectable hope, at least, that the preferred bimetric theory stood a better chance of resolving various puzzles which GR was unable to handle.

Some work has already occurred in this direction. It was mentioned in the last chapter that J.W. Moffat envisioned his own bimetric theory of gravity as a, "preliminary stage for a quantum gravity theory that is not in conflict with some of the basic properties of quantum mechanics" (Moffat, 283). There are, in addition, some problems with string theories which Moffat suggests can be avoided by his bimetric proposal. Specifically, there is a problem producing a description of an eternally expanding universe (the current model as of this writing) that links up consistently with quantum field theory and various string-theoretical structures to which his variable speed of light theory does not succumb (Moffat, 295 – 6).

In this context, it might be appropriate to mention that there has been a considerable amount of speculation regarding possible avenues of dialog between Whitehead's later metaphysical thought and the interpretation of quantum mechanics. Several books have been published recently dealing extensively or exclusively with this subject. However, the treatment of relativity and cosmology in these works is in each case quite limited.

Thus, for example, John Jungerman offers an entire chapter on the subject of relativity, but only to criticize orthodox Einsteinian ideas. Whitehead's alternative theory is given no mention. Indeed, the only connection we find between Whitehead and the phrase "principle of relativity" is in regard to Whitehead's broader metaphysical position (Jungerman 2000, 7, 30 ff). From yet another source, as the title of his book announces, Michael Epperson's recent work is focused on quantum mechanics in the interpretive context of Whitehead's metaphysical system (Epperson 2004). Epperson deals rather more extensively with the connection between relativity and quantum mechanics, but his focus remains on orthodox GR only somewhat contextualized by a consideration of Whitehead's theory of nature (Epperson, 178 ff). However, the only part of the triptych which Epperson mentions is CN, while Jungerman does not touch on this part of Whitehead's work at any point. Finally, Timothy Eastman and Hank Keeton have brought together an important volume of essays from a variety of authors, stemming from the presentations at a conference at the Center for Process Studies in 1998 (Eastman and Keeton 2004, ix – x). However, of the many notable essays in this volume, only one deals specifically with issues stemming from relativity, and that is focused on the problem of time and process in the context of the *special* theory (Eastman and Keeton, 136 – 163). Issues relating to theories of gravity are touched on by a number of authors. However, nowhere in this work is the specifically *cosmological* problem of measurement ever mentioned, nor is its central place in the development of Whitehead's own theory of gravity given any attention. Whitehead's scientific theory of relativity is itself given rather casual attention in only a few spots (Eastman and Keeton, xi, 41, 189), and its relation to other bimetric theories of gravity is not mentioned at all. So while the importance of the contributions to this work are not to be downplayed in any way, their importance to *us* and our problem here is extremely limited.

Thus, we are somewhat on our own here. Returning to, and continuing with our concentration on gravity and cosmology, the primary focus of Moffat's paper (cited above) is on issues relating to the interpretation of redshift data from supernovae. In particular, Moffat observes that the, "observations of supernovae ... at red shifts $0.35 < z <$

1.75 have shown that they appear to be *farther away* and therefore fainter than is to be expected from the standard decelerating model of the universe" (Moffat, 283; my emphasis). This is especially interesting in light of the above observations regarding the QSO in galaxy NGC 7319. The redshift of the QSO, which is a much different cosmological object from a supernova, is vastly greater than the objects Moffat is considering, on the order of z=2.114[136] (Gallianni, *et al*, 88). Still, while it is admittedly enormously speculative to wonder at even the possibility of a connection between these two phenomena, they both share a common problem in that their redshifts make them appear to be further away than other reliable evidence would place them. And such speculation is not contrary to the spirit of philosophical inquiry.[137]

Another factor that might merit consideration – albeit, one that functions at a much more conceptual level – is the massive cosmological asymmetry due to the Abell Cluster bulk flow. It is rather disturbing to contemplate what could cause an enormous chunk of the universe to go marching off to the beat of an unseen drummer in this manner. Whether the causal forces involved be directly gravitational in nature, or something considerably more bizarre like "negative energy,"[138] the disruption to the symmetry of space is astonishing. In case #3, and section III below, we will explore in greater detail the role such symmetry plays in our problem here. But on a purely aesthetic level, this lack of balance is detestable. We have already seen in previous chapters (especially chapter two) how aesthetic factors contribute to the context of theorization, at the very least. What a bimetric – i.e., Whiteheadian – theory offers here is the opportunity to mend this aesthetic imbalance slightly by pushing it out of the very structure of space itself (where it necessarily resides according to GR) and into the realm of the "merely" contingent, physical relations of the cosmos. Now, clearly such a move is devoid of any ultimate, explanatory force. The bulk flow is still there, however we might choose to conceptualize it. However, conceptualizing the flow in a manner that made explicit how a fundamental symmetry can be retained even in the face of such enormous physical asymmetries might yet prove to be of value. So, while the above can scarcely amount to more than a suggestion for a direction of research, it is a suggestion worth making.

Case number two: While Whitehead's criticism of the measurement

---

136. These numbers are simply presented for purposes of relative comparison. We need not explore the details of their interpretation.

137. In email, Moffat expressed cautious interest in the above notion.

138. A fabulously exotic notion which I will not even attempt to explore here.

problem of cosmology stands, no viable Whiteheadian theory of gravity can ultimately withstand the strain of empirical test. Put another way, GR stands unqualifiedly as our fundamental theory of gravity, but the measurement problem of cosmology remains intact. Epistemologically, this would be a catastrophe, for it would mean that cosmology in any genuinely robust sense was impossible. What makes this scenario particularly disturbing is that there is nothing logically incoherent about it. We could, in other words, discover that within the limits of human inquiry, such a scenario just is the case.

Such a result is more than a merely conceptual possibility. It is well within our technological abilities to extend our observational bases substantially beyond the limits of our own solar system. As noted above, the Pioneer spacecraft have already done so. The Voyager 1 probe, though not a viable source of gravimetric data, has in fact gone considerably further into extra-solar space than the Pioneer craft, and is still sending periodic data streams relating to plasma and cosmic particle distributions to Earth. Observational platforms such as the Hubble optical telescope or the Chandra x-ray observatory have vastly increased our ability to detect phenomena that would otherwise be filtered out by the Earth's atmosphere. So, the capacity to greatly extend the range of our cosmological observations beyond our local setting is a well proven fact. As a result, it is an entirely imaginable possibility that we might repeatedly push beyond the limits of our immediate frame of reference and discover that GR remains the only viable theory of gravity available to us. The theoretical possibility is not only there, it is practically imaginable that evidence might be accumulated which consistently defeats any proposed alternatives, and which particularly undercuts any bimetric theories which would otherwise solve the measurement problem of cosmology.

In its starkest formulation, this is clearly not a particularly happy possibility. Cosmology would be reduced to the situation currently facing contemporary string theories – a very extensive body of mathematically clever wild guesses devoid of any meaningful empirical content. Only, in this hypothetical situation, we are projecting that circumstances are so extreme as to preclude any hope of improvement. But such a situation would be quite extreme, which is ultimately what makes this possibility not worth a great deal of concern at this stage of development. The most implausible part of this scenario is that it would ultimately require a refutation not only of any particular known bimetric theory of gravity, but of all such theories in general. It is difficult to imagine this happening any time soon, and not much easier to imagine it happening ever. Experience suggests that it is unwise to underestimate the cleverness of theoreticians, and impossibility proofs are seldom possible outside the context of formal logic and the theory of computable functions. What would be more likely, though only after an extended period of investigation that greatly exceeds anything we have done so far, is that continued failure would discourage any further attempts in such a direction.

There is, however, a third case between the above extremes of a radically alternative theory of gravity on the one hand, and an insoluble measurement problem of cosmology on the other. This is the case in which GR is "true" in a limited sense, but that limitation sharply bounds the context of its application to comparatively – even vanishingly – small regions of space, presumably those in the immediate presence of substantive gravitational bodies. In other words, general relativity is true, but for purposes of cosmological measurement it is essentially reduced to triviality by the vastness of space, where gravitational influences are so relatively inconsequential that the geometry of space is effectively independent of the contingent influences of matter and energy.

There is one aspect about case number three which mandates that we give it a great deal more attention than either of the previous two suggestions: This is, in fact, the direction in which orthodox cosmology is moving. Before we explore how this rather unexpected development is coming about, though, we need to clearly note some of the problems which must be overcome before this can be considered an adequate resolution to the measurement problem of cosmology.

We have previously mentioned the current difficulties encountered when trying to unify micro and macro physics. Even assuming that the formal difficulties of string theories are overcome and, assuming even further, that a formulation is finally produced which offers a meaningful possibility of empirical test, there would remain the logical difficulty of reconciling the mathematical abstractions of such a theory with the extensive realities of human experience. This does not seem like an insoluble problem, though, because at the point where such a unified and empirically meaningful theory is produced, the RNI hypothesis of GR will have been rendered nothing more than a limiting abstraction, along with GR itself. In such a situation, with GR knocked off its current pedestal, the application of Whitehead's principle of extensive abstraction to a unified, string oriented abstract physical theory of the world might conceivably prove to be a simpler program than to directly reconcile experience with the RNI dominated model of GR.

A second problem, also mentioned above, is the Abell cluster bulk flow. GR dominated cosmology could hypothesize a massive super-cluster of galaxies beyond the range of our instruments to account for the so-called "great attractor" bulk flow. But the Abell cluster event is far too big for any such explanation to seem plausible. The problem is that an appeal to an as yet undetected gravitational mass capable of yanking such an incredible volume of galaxies off in a single direction – the only real option available to orthodox GR – would in turn thoroughly disrupt the presumptive symmetry of the cosmos as a whole. This would mean that our presumed solution to the measurement problem (space as a whole is sufficiently uniform to meet Whitehead's requirements) would evidently fall apart. There is an essential connection between symmetry and uniformity which we will explore in more detail in section three. So a breakdown in the symmetry of space is, in turn, a breakdown in the uniformity of space. Because GR identifies the physical characteristics of space with its intrinsic geometry, such a lack of symmetry stands as a massive violation of the presumptive uniformity of space. This is why, in case number one above, the Abell bulk flow was suggested as a case in favor of a bimetric theory of gravity, because it maintains the uniformity of space in the face of the asymmetry of physical relations.

But even in the absence of empirical matters such as the Abell bulk flow, there are theoretical issues of GR which do not lend themselves to being bracketed as mere local effects. In this instance, the issue turns around the necessity (according to the principle of equivalence) of interpreting a point within a gravitational field as essentially the same as a point of, say, a rocket that is undergoing a comparable level of acceleration induced by the rocket's propulsion. But the rocket's acceleration, unlike a gravitational field, is not a distortion of the actual geometry of space. As such, it seems like it should be possible to redefine the frame of reference around the rocket so that it stands within an inertial frame, betraying a fundamental difference between it and a true gravitational field. To avoid this problem, the formalism introduces "pseudo" gravitational fields to make everything balance. This aspect of the equivalence principle has troubled a number of thinkers, to the extent that, in older texts at least, some effort was commonly made for a kind of apology for it. For example, D.F. Lawden wrote:

> However, the reader is probably still not convinced that the distinction between accelerated and inertial frames has been effectively eliminated, but only that it has been concealed by means of a mathematical device having no physical significance. Thus, he may point out that the gravitational fields which have been introduced to account for the inertial

> forces are 'fictitious' fields, which may be completely removed by choosing an inertial frame for reference purposes, whereas 'real' fields, such as those due to the earth and sun, cannot be so removed.... These objections may be met by attributing such 'fictitious' fields to the motions of distant masses within the universe. ... On this view, therefore, inertial frames possess particularly simple properties only because of their special relationship to the distribution of mass within the universe (Lawden, 129).[139]

The camel we are being asked to swallow here is that the possibility of determining if a frame is accelerated or not is made real by virtue of the influences of the rest of the cosmos on the point in question. There is a certain necessity to swallowing it, however, within the framework of GR. Absent an independent geometrical frame to which motion can be compared as either inertial or accelerated, then the only space in which and to which such comparison can be made is the one whose subsistent ontological reality is in turn referred to the contingent "distribution of mass within the universe" that GR gives us. Whitehead saw this as especially troublesome in the context of rotational phenomena, such as the equatorial bulge of the earth (R 87 – 8).

It is very difficult to find a satisfactory discussion that attempts to resolve these sorts of local/global frames of reference issues in the physics literature. Unfortunately, this literature is not indexed in a particularly thorough manner, and it is thus unreasonable to conclude that lack of evidence is the same as evidence of lack. But none of the proposed solutions seem very happy. Are we to conclude that the equatorial bulge of the earth is caused by distant galaxies? Then we are obliged to suppose that it would be different if that distribution were fundamentally other than it is? On the other hand, we can calculate measurable tidal effects on the earth, were the entire mass of the galaxy concentrated in a point at the center of our galaxy (see chapter five, section three, above). If we are to take Whitehead's argument for the relationality of nature seriously, can we exclude the above possibility? "A scientific object such as a definite electron is a systematic correlation of the characters of all events throughout nature" (CN 158). Nature, Whitehead has told us, is a systematic whole. What precludes that systematicity from causing the earth to bulge at the equator solely on account of the distributions of gravitating

139. Although it was updated in 1981, republished in 1982, and later picked up by Dover Publications in 2002, Lawden's text was originally written in 1961. So, while it is a usefully compact treatment of the subject which brings together a number of important topics within its covers, this must be counted as an older text.

masses in the universe, which of course includes the very distant galaxies?

Ultimately, the point to be understood here is simply that, if we mean to incorporate GR into a cosmological framework which nevertheless avoids the measurement problem, we are faced with some tricky interpretive issues. We must explain how it is that, on the one hand GR's geometrical contingencies are fundamentally unimportant when the universe is considered at large, but on the other hand it is these very relationships that determine the nature of an inertial frame. While these are problems that certainly deserve to be addressed, the above should not be viewed as insurmountable objections. As already suggested in the above, one approach would be to embrace the poles of the dilemma as mutually explanatory: it is *precisely* the relational character of the distribution of matter in the universe which leads to the relative triviality of the geometric contingencies of GR at cosmological scales, and which is also the central contributing factor in the structure of inertial frames on a local scale.[140] So, while we must not forget these problems, let us at least bracket them for the time being, and consider instead what sorts of things would have to be the case in order for a reconciliation to occur here between Whitehead and Einstein.

Whitehead himself at one point suggests such a possibility. Whitehead says that, "If the above (gravitational) formula gives results which are discrepant with observation, it would be quite possible with my general theory of nature to adopt Einstein's formula, based upon his differential equations, for the determination of the gravitational field" (R 84). Unfortunately, Whitehead offers no clue as to how this adoption could occur within his theory of nature. Palter also notes this statement of Whitehead's, and briefly examines it in the context of the possibility (suggested by Eddington), that in empty space, the characteristics of ten essential functions of the Riemann curvature tensor all vanish, while the other ten vary arbitrarily (see also CN 182 – 3).[141] Palter rejects this suggestion on the ground that it is too purely formal a solution to the problem. Instead, Palter notes that Einstein's metric tensor $g_{\mu\nu}$ can be split into two parts, a fixed, inertial tensor, and a variable gravitational tensor, and suggests that this is the possibility which Whitehead had in mind

---

140. Although, the Abell bulk flow seems like a much more difficult problem in such an interpretation.

141. This is an aspect of the Riemann tensor we have not explored. Briefly, because of various symmetry conditions, there are only 20 significant degrees of freedom in the Riemann tensor, even though with four indices there are mathematically a total of 256 such functional possibilities.

(Palter, 207 – 208). We will consider a third possibility, from which Eddington's suggestion emerges as a natural consequence, but which Whitehead himself may never have considered.

Before turning to this question, which will occupy the bulk of section 3, let us first review what we have learned so far about Whitehead's uniformity criterion. The purpose of uniformity is, of course, so that we can legitimately apply our physical theories to events that are taking place well beyond the range of our immediate experience. Since we cannot be physically present at such distant events, we must be in possession of systematically effective relations that are at once empirically adequate to the task of logically translating our measurements to and from the events in question. We cannot carry our yardstick to the events, so we must somehow be able to logically project our yardstick to where we ourselves are unable to go. Whitehead argued not only that there must be a system of uniform relations in place so that we can make such a logical move, but in a radically empiricist gesture, he went on to argue that the uniformity of such relational structures is itself a content of our experience. From these considerations, Whitehead went on to construct a theory of gravity that was an alternative to general relativity. This theory was bimetric in nature, employing both a varying metric $J_{\mu\nu}$ for the contingent physical and gravitational relations, and a Euclidean metric $G_{\mu\nu}$ for the underlying uniformities of pure space. The physical and material phenomena emerge from the interaction, which can be quite layered and complex, of these two metrics.

However, there are a number of things which Whitehead's uniformity criterion is *not*. It is not, in the first place, an absolute commitment to a Euclidean, or even a flat, metric for the uniform relations of space. As he explicitly stated, Whitehead used a Euclidean structure in his own theory purely on account of its formal simplicity. But there are other kinds of uniform geometry which work equally well within the general framework of Whitehead's philosophical position, and if the evidence were to suggest that a uniformly spherical or uniformly hyperbolic space better accounted for the phenomena, then his criterion would also work perfectly well with such geometries. Whitehead's comparatively sanguine attitude toward the ultimate mathematical basis of his theory, and his lack of attention to it once it was published, all show rather convincingly that he viewed his theory as an extended *exempli gratia*, and nothing more. It is therefore interesting to note that of the considerable range of bimetric theories that have been proposed since Whitehead's original publication of R, most if not all of them have been built around a flat metric. Whitehead was certainly cognizant of the work on uniform spaces – also known as spaces of constant curvature – which was a vigorous area of research when he was coming up through University. But no effort has been expended in generalizing Whitehead's theory to such spaces except for a few very early articles by G. Temple. Since there are a number of bimetric theories besides Whitehead's which are indisputably viable, pushing such a generalization is of little more than formal interest at this time.

Another thing that Whitehead's criterion does not commit us to is any sort of *a priori* knowledge regarding the geometric structures of space. While it is the case, according to Whitehead, that the uniformity of space is an immediate content of experience, there is no reason to suppose that the nature and structure of that uniformity can ever be had without extensive investigation. *That* space is uniform is given to us; *how* space is uniform is not.

Related to this point is the question of whether Whitehead's theory itself commits us to a "prior geometry." Clifford Will characterizes a prior geometric theory as, "any metric theory that contains 'absolute elements,' fields or equations whose structure and evolution are given *a priori* and are independent of the structure and evolution of the other fields of the theory," (Will 1993a, 79). But as we have already noted, the geometry in Whitehead (and, frankly, most other bimetric theories) is not given *a priori* in the sense of prior to experience or investigation. Nor is it clear in what sense the geometry in Whitehead's theory can be called "absolute."

On this last point, consider that what uniformity requires is constant spatial curvature. This means only that, at every point of the space under investigation, the curvature relations of the space are the same. If all one knows about any such point is how space is curved there, one effectively has no idea of where in the space that point might be. But keep in mind, these are specifically *spatial* relations. In PNK, Whitehead opens an interesting door, whose implications he never fully explored. Because of its importance, I quote Whitehead here at length:

> There are three different types of meaning which can be given to the idea of 'space' in connection with external nature. (i) There is the four-dimensional space of which event-particles are the points and the rects and point-tracks and null-tracks are the straight lines. *In the geometry of this space there is a lack of uniformity between the congruence theories for rects and for point-tracks.* (ii) There are the three-dimensional momentary (instantaneous) spaces in the moments of any time-system *a*, of which event-particles are the points and rects are the straight lines. The observed space of ordinary perception is an approximation to this exact concept. (iii) There is the time-less three-dimensional space of the

> time-system a, of which point-tracks are the points and matrices[142] include the straight lines. This is the space of physical science (PNK 137 – 8, my emphasis).

The picture Whitehead gives us is of a hierarchy of spaces, each of superior uniformity to that which preceded it, until one is at last at the maximally uniform space of physical science. And here we must recognize that the ideally uniform space is just and only *space*; it is three-dimensional and not, on that account, space*time*. In the larger, four-dimensional space of event-particles (which are themselves idealizations from genuine events) there are a variety of candidates for the role of "straight line," but no uniform method of relating them.

The door this opens is this: space is uniform, but not fixed. In other words, we have the possibility here that the uniformity of space is itself something which can evolve – perhaps uniformly – in the process of nature. I deliberately avoid the saying "evolve with time" because, as we have already seen, in Whitehead's scheme of things, space*s* and time*s* are themselves emergent characters from the flow of events. Indeed, each individual space of some given time system receives its ordering from its intersections with other time systems. This ordering is uniform and constant – *within a given time system* – but there is nothing to suggest that, as new time systems emerge from the flow of events, the transition from one uniformity to the next should be as simple as a mere identity. This, again, is not a possibility that we see explicitly discussed anywhere within Whitehead's corpus. However, we find in his later, metaphysical works, the hint at least of the idea that the laws of nature may indeed evolve from one epoch to another in his frequent, pointed references to the peculiar characteristics of *this* cosmic epoch (PR, 66 ff, 91 ff, 98, etc.). With this background in hand, we shall now turn to the connection between uniformity, symmetry, and space.

III. Maximally Symmetric Spaces:

We have previously noted the connection between spaces of constant curvature and Whitehead's uniformity criterion. We now need to expand a little further on these ideas, noting in particular the developments which occurred shortly after the publication of R and which have come to contribute directly to contemporary cosmology. However, as one might expect, before we can move forward, we first need to step back.

---

[142]. This is a Whiteheadian use of the term "matrix," and is quite different from how we have used the term in previous chapters.

While the study of spaces of constant curvature had been a topic of active interest at the time Whitehead was studying mathematics at University, little additional progress had occurred since the days of Klein, Killing, and Clifford, and things had not advanced significantly in the first two decades of the 20th Century up to and including the time when Whitehead published R. Ideas were scattered throughout the literature, there was little uniformity in the use of symbols to communicate a common idea, etc. (Epple, 942). In addition, a variety of classificatory problems presented themselves which, in part because of the previously mentioned lack of a common set of symbols, were proving to be difficult if not intractable. However, one thing was known with great confidence even at the time of Whitehead's *Universal Algebra*: the principal tool for tackling these problems was the mathematical theory of groups. Although we have examined some of the formal characteristics of groups in previous chapters, it will serve us well here to reiterate and expand on some of that discussion.

To begin with, while it is certainly the case that mathematical groups,[143] like any other formal structures in mathematics, can be studied on their own account,[144] they are also of special use in the study of, and application to, other mathematical structures as well. Thus, the elements of a group can operate on some other object, such as the member of a set, and transform that element of the set into another element of the set in some specifiably precise manner. Thus, if Γ is our group, and $\gamma_i \in \Gamma$ are the elements of the group, while $x_j \in X$ are the elements of the set X, then it is also the case that $(\gamma_i x_j) \in X$. Because Γ is a group, each such transformation has a unique inverse. Which is to say, for every specific $\gamma_i \in \Gamma$ there is exactly one $\gamma_i^{-1} \in \Gamma$ which will undo γ_i's action on x_j, and return us to the original element of X with which we started. Because the group elements each have a unique inverse, there is also a unique identity element which performs the null transformation on the objects of X, "transforming" each such object into itself. Suppose now that the structure our group is operating on is not some arbitrary set, but a geometrical space. In this case, the objects being transformed the one into the other are not just elements, they are geometrical points. What are some of the features of this new

143. From here on I will simply refer to these structures as "groups," and not add the additional qualifiers of "mathematical" or "formal." I will not use the word "group" in any other way than this.

144. In which case it is considered a part of "abstract" algebra, as opposed to the more "concrete" manifestations discussed here. This is a distinction which is likely to cause a non-mathematician to say, "go figure."

arrangement?

Well, one can, of course, stipulate any imaginable transformation one likes, and so long as each such change is uniquely undo-able, one will stay within the basic rules of a transformation group. However, the only really interesting such groups of transformations are those which retain some mathematically interesting aspect of the original geometric space. What is of interest is not fixed at the outset, but determined by the nature of one's inquiry. But the sorts of interesting things one can work with order themselves rather naturally into a hierarchy of increasing specificity. It was Felix Klein who first clearly understood the possibilities of such a classificatory scheme predicated upon the interesting group theoretical transformations of geometric spaces. Klein articulated what has since come to be known as his "Erlanger Program"[145] in 1872 for the classification of geometries according to the group structured transformations which could be applied to each type of space while keeping various relevant levels of spatial structure intact. Thus, one type of transformation preserved only the "neighborhoods" of a point; if a point started out "near" another point – in some deliberately vague sense – then the transformation of that point and the space itself only saved that vague sense of nearness. Such structures are studied under that branch of mathematics that later came to be known as point-set topology. Other transformations preserved not only nearness, but projective relations as well, giving us projective geometry. Other, more rigidly limited groups of transformations preserved not only nearness and projectivity, but actual metric relations themselves, thus specifying metric geometry.[146]

There is a variety of ways in which this method of classification can be extended and made more subtle. What is of importance for us, though, is that this same technique serves to classify spaces of constant curvature. It is scarcely an accident that the man who formulated the Erlanger program was one of the leading investigators into the nature of such spaces. Consider that one of the structure preserving group transformations that can be applied to a space is one that preserves the curvature relations of any given point of that space. However, the only way this can happen in a way that conforms to the rules of group theory is if the curvature relations (or, at least, the "interesting" curvature relations) of any one point are the same for *all* points. If this were not the case, then the group

145. Klein first described it in a speech given upon his admission to the faculty of the University of Erlangen.

146. See, for instance, (Kline, 917 – 20) or (Kramer, 401 – 433) for a detailed discussion of Klein and his project.

transform of one point's structure to another's would lose that structure, which we've previously stipulated as interesting. So a space of constant curvature is one in which the curvature relations of the space can be shown to be strictly analogous to one another by their mutual inter-translatability through the rules of the relevant group.[147]

There is a close connection between spaces of constant curvature, Riemannian geometries in particular, and differential geometry in general, which is readily evident the moment one considers the intrinsic focus of these latter on curvature. Consequently, the study of spaces of constant curvature amounts to the application of a particular class of group theoretical transformations to the very same geometrical structures which form the foundation of general relativity and, indeed, all theories of space and gravity. The final puzzle piece that ties all of this in with our considerations here is the recollection that groups are, along with all of their other properties, inherently about symmetry. Symmetry, again, is a kind of sameness. An animal or a person exhibits a kind of bilateral symmetry because it is the same on the left side as on the right. A geometrical figure like a square shows a more robust group of symmetries, because it can be flipped and or rotated around a variety of axes and still appear the same, and so on. A space of constant curvature is highly symmetrical because it displays a thoroughgoing measure of sameness – the constancy of its curvature – at each point, and viewed from a variety of angles. Because the possibility of measurement requires such constancy in order for there to be knowable congruence relations, it is evident that Whitehead's uniformity criterion is in essence a requirement for symmetry.

But, as we have seen, symmetries come in a variety of flavors, and even in the context of spaces of constant curvature, these varieties had not been fully explored at the time that Whitehead was writing his triptych. Two events occurred in 1925 and 1926 which contributed significantly to this subject. The first was the work by Heinz Hopf whose dissertation finally unified and modernized the ideas and symbolism of spaces of constant curvature. As Epple puts it, Hopf "presented a completely revised treatment of the problem of (Clifford-Klein) space forms that made it superfluous to look into the older literature any more," (Epple, 942). Shortly after Hopf's work, in 1926 Elie Cartan published a pair of papers that initiated the study of a very particular type of such spaces, the *maximally symmetric spaces* (Helgason 1962, vii).[148] What made Cartan's

147. (Wolf 1984) is the standard reference on this subject.

148. Helgason's volume is the standard reference on this subject. Note, however, that in later editions, the very brief historical remarks in the preface have been removed in

work of such importance was that he brought together two types of mathematical structure which had, until then, been developed independently of one another. The first, of course, was our spaces of constant curvature. The second, though, was the notion of a Lie group, a system of continuous transformations that nevertheless obey the rules of a group.[149] It is this *continuum* of allowable transformations that allowed Cartan to develop the maximum possible level of symmetries of these spaces. Finally, it turns out that the best known examples of symmetric spaces are the Euclidean, spherical, and hyperbolic spaces (Wolf, 231).[150] These, of course, are precisely the kinds of space which Whitehead explicitly mentions in the opening pages of R. So, it is ironic – at least – that the mathematical tools that could fully characterize Whitehead's uniformity criterion appeared only after he presented his uniformity argument, and in a language (French) in which he had no proficiency at all (Lowe 1990, 15).[151]

This is the last piece needed to outline our possible reconciliation between Whitehead and Einstein. The thread that ties everything together comes to us from cosmology itself. We are told that, "observationally we know that the universe is homogenous and isotropic in space," which is important because, "homogeneity and isotropy ... imply that a space is *maximally symmetric*," (Carroll, 324, my emphasis). And there we have it, handed to us on a platter. Any maximally symmetric space easily qualifies as meeting Whitehead's uniformity criterion, and now we are informed that, as a matter of observed fact, this is exactly what we see on a cosmic scale. Now, it must be immediately and emphatically added here that it is only *space* that is maximally symmetric, not *spacetime* as a whole (*ibid*).

---

favor of a discussion of some comments regarding the earlier editions of the book. Cartan's 1926 papers, and others he contributed to the subject, are listed in Helgason's bibliography, and can be found in Cartan's *Oeuvre Completes*.

149. Making Lie groups work in this situation is rather tricky. Because Lie groups are so much more intimately embedded in the umbrella structure of Lie algebras, it is necessary to create a function that will translate any given point and its neighboring space to the zero point of the given coordinate system, so that the Lie algebraic structures can be allowed to function properly and, by default then, the Lie group structures as well.

150. These are not the only maximally symmetric spaces, but they are the only ones that are locally Euclidean in nature (Carroll, 143). For this reason, they are the only such spaces that are viable models of cosmology.

151. Conversely, Cartan published only one paper in English, and that was a collaborative effort. An examination of Cartan's published papers between 1920 and 1927 shows that he never even cites an English language work.

However, even in the context of spacetime, maximally symmetric models are useful approximations to a kind of "ground state" description of a "pre-universe" (my term) that is not yet suffused with matter and energy (Carroll, 324 ff). And, as we have seen above, while Whitehead's theory did not explicitly anticipate such a possibility, neither does it explicitly rule it out. Indeed, the quote from PNK given above rather seems to invite us to entertain just such an interpretation of Whitehead's criterion as would serve to resolve our problem here.

One might almost suppose that, after all this work, it is almost as if cosmology just sold the pass to Whitehead. But the effort expended in reaching this point has hardly been gratuitous, nor is the matter as settled as the above statement by Carroll might seem to leave things. To begin with, connecting the uniformity criterion with maximally symmetric spaces requires that we first appreciate the nature and depth of the measurement problem of cosmology, and the kinds of obstacles toward the resolution of that problem thrown in our path by orthodox GR. We have to accept an interpretation of GR as somehow both spectacularly true, but largely trivial. Moreover, it is worth noting in this regard that the idea of space as maximally symmetric is itself seldom given extensive treatment in works on cosmology. Stephen Weinberg devotes a chapter to the subject in his text from 1972, cited in (Carroll, 492). Carroll himself devotes several pages to the topic (Carroll, 139 – 144, 323 – 9). But these are more the exceptions than the rule. Thus, (Misner, Thorne, Wheeler 1973) does not mention the subject anywhere, nor does (Lawden 2002), although both give rather brief mentions to the more general idea of spaces of constant curvature. In this regard, it is interesting to note that Lawden's text was originally published in 1961, a year before Helgason unified and standardized the topic of maximally symmetric spaces.

This proposed reconciliation is hardly unproblematic. For one thing, Whitehead's argument was that his uniformity criterion was at least an epistemologically necessary condition of scientific knowledge of the cosmos, while Carroll tells us it is merely an observed fact of the contingent distributions of matter and energy that such uniformity is found in the universe. Moreover, it is a fact which only happens to be true *now* – at earlier periods of the universe, this contemporary uniformity – which is due to the general scarcity of matter and energy in comparison to the vastness of space – was most definitely not the case (Carroll, 328 – 9). Being lucky is not the same as being right. The fact that we just happen to find ourselves within a cosmic epoch which is coincidentally structured in such a manner as to permit of a meaningful scientific cosmology is not very intellectually satisfying as an explanation. It is rather as if we've stepped into a Sid Harris cartoon, in which the middle step of our problem has been resolved because, once again, a miracle occurred.

Yet another problem stems from the fact that, if the uniformity found depends only upon empirical evidence, then empirical evidence can snatch it away just as easily. It is here that the Abell cluster bulk flow problem really strikes home. A 600,000,000 light year radius chunk of the universe being yanked out of place by a massive (presumably) gravitational tug speaks to a substantial breakdown of symmetry at a cosmological scale. This is why we made note on several occasions earlier in this chapter of this phenomenon, especially in the context of symmetry. Uniformity just *is* symmetry, and a loss of the latter is a loss of the former. And because GR eliminates the distinction between physics and geometry, this asymmetry cannot be ignored by any theory of gravity that is built around the former, yet hopes to escape the measurement problem of cosmology. For orthodox GR, the Abell bulk flow is not simply an as yet unexplained physical phenomenon, it is a structural feature of space itself. As such, so is the inherent asymmetry and massive failure of uniformity associated with it.

IV Conclusion: Looking Back and Looking Ahead:

Since we have already summarized the first five chapters at the beginning of this one, there is no need to rehearse that discussion again. Rather, let us bring together the last of the threads of our argument, and from that position glance at some of the directions this research could be continued beyond the bounds of this investigation.

Clearly it is the case that a maximally symmetric space entirely captures the formal and structural features Whitehead saw as necessary for any adequate background to a scientific theory of space and gravity. Only such a background would possess the required level of uniform logical connection that would permit one to project the requisite congruence relations beyond the limits of our direct physical encounters, and thus give meaning to cosmological measurements. This is actually a very satisfying result to have in hand, for it gives us an exceptionally robust characterization of Whitehead's uniformity criterion in a well developed format that lends itself to a variety of applications. This is a point Helgason makes with respect to the purely mathematical properties of these symmetric spaces (Helgason, vii). But the possibilities here go beyond just mathematics. This formalization of Whitehead's criterion not only provides us with a rigorous body of thought with which we can evaluate cosmological theories, but one which reflects back upon Whitehead's criterion itself and gives it the considerable weight of a well developed discipline. Whether through the direct path of its own evolution, as with the previously examined case of Whitehead's theory of extension, or in this case here which has emerged through the happy conjunction of two lines of thought which had advanced at least somewhat independently, the fact that aspects of Whitehead's argument lend themselves so well to rigorous development stands powerfully in their favor.

Furthermore, there are two lines of research which suggest themselves as paths for resolving the measurement problem of cosmology. The one which we have just considered has the advantage of being the most conservative line to follow. Reconciling Whitehead's criticisms of orthodox GR with contemporary cosmology along the lines described above in section three requires no major revisions or revolutions of operational theoretical systems. We get to keep well developed theories and continue existing lines of research. Moreover, since it is clear that major compromises in the interpretation of GR must occur in order to link it with micro physics, the way is cleared to applying Whitehead's larger philosophy of nature to cosmology. GR is necessarily knocked off the pedestal of "The Way The World Is" by such developments as string theory, and relegated to the status of a very good approximation of reality by the work of physics itself. However, there are several clouds on the horizon for this approach. Quite aside from the directly relevant issues of things such as the Abell cluster bulk flow, other empirical problems (the pioneer anomaly, the QSO in NGC7319) are not, as of this writing, accounted for within existing theory.

The other path,[152] of moving into a genuinely Whiteheadian bimetric theory of gravity, while fairly revolutionary, is nevertheless not without promise. As we have seen, there is reason to hope that bimetric theories resolve some issues that stand in the way of linking the theory of gravity with a functional string theory, and there are redshift phenomena of supernovae that are suggestively analogous to the QSO in NGC7319 phenomenon. Moreover, while there is no reason to suppose that a bimetric theory would make accounting for the Abell bulk flow simpler, at least such a theory would separate the contingent relations of physics from the necessary uniformities of geometry. Cosmology, in such a scheme, continues to be at least possible. The weakness of this approach is that it has not enjoyed the depth and degree of investigation one finds in orthodox GR. Consequently, it is not as well developed, and the possibility that such

152. And one which, for the record, this author finds far more attractive.

theories have even more profound conflicts with the empirical evidence cannot be ruled out. On the other hand, if the physics community can be persuaded to move beyond the mis-classification of Whitehead's own theory of relativity as "quasi-linear," then it will be possible to see how Whitehead's thoroughgoing development of a philosophy of nature and a robust theory not only of measurement but of abstraction as well, powerfully motivates an intensification of inquiry into these bimetric theories. At the same time, if the philosophical community came to be more generally aware of the existence of viable bimetric theories of gravity on the one hand, and mereotopologies which directly trace their genealogies to Whitehead's theory of extension on the other, then these vindications of his philosophical positions could significantly broaden interest in the Whiteheadian approach to nature, abstraction, space, time, gravity, etc., thus contributing to the dialog on all of these topics.

Moreover, we have here the opening up of a much broader field of application of Whitehead's criterion. Though he focused his critiques on general relativity, Whitehead's discussions of measurement and uniformity are not logically bound to cosmology alone. So while such a program runs far beyond the bounds of our specific investigation, we should acknowledge the fact that we can use Whitehead's ideas to examine theories of measurement *in general*. For any act of measurement must live up to the uniformity criterion in some form or other. This is true not only of the physical sciences, where such measurement typically possesses a more or less literally spatial component, but to any science and any form of measurement whatsoever. Even the measures used in the social sciences, based as they are upon statistical samplings, require a kind of uniformity in order for statistical tools to be able to extract valid information from the noise of the sample. We even find the cluster of phenomena from which such samplings are derived referred to as a "sample space." Thus, a comparative study of the kinds of uniformities employed by different sciences veritably cries out for investigation.

Finally, the particular vindication of Whitehead's thought in the context of cosmology and macro physics directly contributes to the task of unifying physics at the cosmological and the quantum levels. As we have already noted, there is a vigorous and ongoing line of research which applies Whitehead's philosophical ideas to the interpretation of quantum mechanics. What we now see is that Whitehead's thought continues to provide us with an equally vigorous interpretation of macro physics as well. We thus have a single philosophical framework with which we can interpret both. For let us be frank: any interpretive scheme which contributed only to the understanding of one, but offered nothing to the other, could ultimately contribute to neither. Such a profoundly limited interpretation would wreck itself on the shoals of the two levels of physics.

Happily, that is not our situation here. So it is upon this note that we shall close our inquiry.

REFERENCES

Anderson, John D., *et al.* 2004. "Study of the Anomalous Acceleration of Pioneer 10 and 11," *Physical Review D*, Vol. 65, No. 8, 082004, pp. 1-50.

Birkhoff, Garrett and MacLane, Saunders. 1997. *A Survey of Modern Algebra*, A. K. Peters, Wellesley. (Corrected reprint of the fourth edition.)

Bishop, Richard L. and Godlberg, Samuel I. 1980. *Tensor Analysis on Manifolds*, Dover Publications, Inc., NY.

Carroll, Sean. 2004. *Spacetime and Geometry: an Introduction to General Relativity*, Pearson Education Inc. publishing as Addison Wesley, San Francisco.

Clarke, Bowman L. 1981. "A Calculus of Individuals Based on 'Connection'," *Notre Dame Journal of Formal Logic* volume 22, number 3, pp 204 – 218. July.

---. 1985. "Individuals and Points," *Notre Dame Journal of Formal Logic* volume 26, number 1, pp 61 – 75. January.

Code, Murray. 1985. *Order and Organism*, SUNY Press, Albany.

Collins, Harry and Pinch, Trevor. 1993. *The Golem: What everyone should know about science*, Cambridge University Press, Cambridge.

Davis, Philip J., and Hersh, Reuben. 1981. *The Mathematical Experience*, Houghton Mifflin Co., Boston.

de Laguna, Theodore. 1922a. "The Nature of Space—I," *The Journal of Philosophy*, Vol. 19, No. 15. (Jul. 20, 1922), pp. 393-407.

---. 1922b. "The Nature of Space—II," *The Journal of Philosophy*, Vol. 19, No. 16. (Aug. 3, 1922), pp. 421-440.

---. 1922c. "Point, Line, and Surface, as Sets of Solids," *The Journal of Philosophy*, Vol. 19, No. 17. (Aug. 17, 1922), pp. 449-461.

Dewey, John. 1991. *Logic: The Theory of Inquiry*, vol. 12 of The Later Works, 1925 – 1953, edited by Jo Ann Boydston, Southern Illinois University Press, Carbondale.

Dirac, P. A. M. 1996. *General Theory of Relativity*, Princeton University Press, Princeton.

Dressler, Alan. 1994. *Voyage to the Great Attractor*, Alfred Knopf, NY.

Eastman, Timothy E., and Keeton, Hank. 2004. *Physics and Whitehead: Quantum, Process and Experience*, State University of New York Press, NY.

Eddington, Arthur S. 1924. "A Comparison of Whitehead's and Einstein's Fomulae," *Nature*, (Feb. 9, 1924), pg. 192.

Einstein, Albert. 1961. *Relativity, the Special and the General Theory*, Three Rivers Press, NY.

Einstein, Albert, *et al*. 1952. *The Principle of Relativity,* Dover Publications, Inc., NY.

Epperson, Michael. 2004. *Quantum Mechanics and the Philosophy of Alfred North Whitehead*, Fordham University Press, NY.

Epple, Moritz. 2002. "From Quaternions to Cosmology: Spaces of Constant Curvature, ca. 1873–1925," *Proceedings of the ICM*, vol. 3, 935—946, from the International Congress of Mathematicians, Beijing, August 20 - 28, 2002.

Freedman, Wendy L., *et al*. 2001. "Final Results from the Hubble Space Telescope Key Project to Measure the Hubble Constant," *The Astrophysical Journal*, vol. 553, 20 May 2001, pp. 47-72.

Fowler, Dean. 1974. "Disconfirmation of Whitehead's Relativity Theory – A Critical Reply," *Process Studies*, vol. 4, no. 4, winter 1974.

---. 1975. *Relativity Physics and the Doctrine of God: a Comparative Study of Einstein and Whitehead*, Claremont Graduate School, 1975. University Microfilms no. 76-15,755.

Gallianni, Pasquale, *et al*. 2005. "The Discovery of a High-Redshift X-Ray–Emitting QSO Very Close to the Nucleus of NGC 7319," *The Astrophysical Journal*, vol. 620, 10 February 2005, pp. 88–94.

Greene, Brian. 2000. *The Elegant Universe*, Vintage, NY.

Griffin, David R. 1986. Editor, *Physics and the Ultimate Significance of Time*, SUNY Press, Albany.

Grünbaum, Adolf. 1963. *Philosophical Problems of Space and Time*, Alfred A. Knopf, inc., NY.

Helgason, Sigurdur. 1964. *Differential Geometry and Symmetric Spaces*, 2nd edition, Academic Press, NY.

Hersch, Reuben. 1997. *What is Mathematics Really?*, Oxford University Press, Inc., NY.

Hilbert, David. 1915. "Die Grundlagen der Physik. (Erste Mitteilung)" Nachrichten von der Gesellschaft der Wissenschaftern zu Göttingen. Mathematisch-phyikaische Klasse (1915), 395-407.

Hintikka, Jaakko. 1999. *Inquiry as inquiry : a Logic of Scientific Discovery*; selected papers, v. 5; Kluwer Academic Publishers, Dordrecht / Boston.

---. 2001. "Is Logic the Key to All Good Reasoning?" *Argumentation* 15: 35–57. Kluwer Academic Publishers.

Holton, Gerald. 1980. *Thematic Origins of Scientific Thought*, Harvard University Press, Cambridge.

Jungerman, John A. 2000. *World in Process*, State University of New York Press, NY.

Kennicutt, Robert C., Jr., *et al.* 1995. "Measuring the Hubble Constant with the Hubble Space Telescope," *The Astronomical Journal*, Vol. 110, No. 4, October 1995, pp. 1476-1491.

Kline, Morris. 1972. *Mathematical Thought from Ancient to Modern Times*, vol. 1 – 3, Oxford University Press, NY.

Kramer, Edna. 1980. *The Nature and Growth of Modern Mathematics*, Princeton University Press, Princeton.

Kraus, Elizabeth M. 1998. *The Metaphysics of Experience*, Fordham University Press, NY

Lawden, D.F. 2002. *Introduction to Tensor Calculus, Relativity and Cosmology*, Dover Publications, Inc., NY.

Lowe, Victor. 1975. "A. N. Whitehead on his Mathematical Goals: a Letter of 1912," *Annals of Science*, 32 (1975), 85 – 101.

---. 1985. *Alfred North Whitehead: The Man and his Work. Volume I: 1861 – 1910*, the Johns Hopkins University Press, Baltimore.

---. 1990. *Alfred North Whitehead: The Man and his Work. Volume II: 1910 – 1947*, the Johns Hopkins University Press, Baltimore.

McKeon, Richard. 1941. *The Basic Works of Aristotle*, Random House, Inc., NY.

Misner, Charles W., Thorne, Kip S., Wheeler, John Archibald. 1973. *Gravitation*, W. H. Freeman and Co., San Francisco.

Moffat, J. W. 2003. "Bimetric Gravity Theory, Varying Speed of Light and the Dimming of Super-novae," *International Journal of Modern Physics D* Vol. 12, No. 2 (2003) 281–298.

Palter, Robert. 1960. *Whitehead's Philosophy of Science*, The University of Chicago Press, Chicago.

Piso, M.I., Ionescu-Pallas, N., Onofrei, S. 1994. "Linear Bimetric Gravitation Theory," *eprint arXiv:gr-qc/9407017*, http://arxiv.org/abs/gr-qc/9407017, July 14, 1994.

Poincaré, Henri. 1913. "La Mesure du Temps", Rev. Metaphys. Morale 6, 1 (1898) Available in English as Chapter II of *The Value of Science*, George Bruce Halsted, Translator, Science Press, New York.

Polya, George. 1990. *Mathematics and Plausible Reasoning*, volume 1, 12th printing, Princeton Paperbacks, Princeton, NJ.

Pratt, Ian and Schoop, Dominik. 1998. "A Complete Axiom System for Polygonal Mereotopology of the Real Plane," Journal of Philosophical Logic 27: 621–658.

Randell, D. A., Cui, Z. and Cohn, A. G. 1992. "A spatial logic based on regions and connection," in B. Nebel, C. Rich, and W. Swartout (eds), *Principles of Knowledge Representation and Reasoning: Proceedings of the Third International Conference* (KR '92), Morgan Kaufmann, Los Altos, CA, 1992, pp. 165–176.

Rayner, C. B. 1954. "The Application of the Whitehead Theory of Relativity to Non-static, Spherically Symmetrical Systems," *Proceedings of the Royal Society of London. Series A, Mathematical and Physical Sciences*, Vol. 222, no. 1151 (Mar. 23, 1954), pp.s 509 - 526.

Reichenbach, Hans. 1957. *The Philosophy of Space and Time*, Dover Publications, Inc., NY.

Rosen, Nathan. 1940. "General Relativity and Flat Space I & II," *Physical Review*, vol. 57, January 15, 1940, pp. 147 – 153.

---. 1974. "A Theory of Gravitation," *Annals of Physics*, vol 84, 1974, pp. 455 – 473.

Rothstein, Edward. 1995. *Emblems of Mind: the Inner Life of Music and Mathematics*, Avon Books, NY.

Schild, A. 1956. "On Gravitational Theories of Whitehead's Type," *Proceedings of the Royal Society of London. Series A, Mathematical and Physical Sciences*, Vol. 235, no. 1201 (Apr. 24, 1956), pp.s 202 – 209.

---. 1961. "Gravitational Theories of the Whitehead Type and the Principle of Equivalence," *Proceedings of the International School of Physics, Enrico Fermi course 20*, edited by C. Møller, Academic Press, New York, pp. 69 – 115.

Schilpp, Paul Aurther. 1951. Editor, *The Philosophy of Alfred North Whitehead*, Tudor Publishing Company, NY.

Seaman, Francis. 1955. "Whitehead and Relativity," *Philosophy of Science*, vol. 22, no. 3, July 1955, pp. 222 - 226.

---. 1975. "Note on Whitehead and the Order of Nature," *Process Studies*, Summer 1975; vol. 5: 129-133

Strauss, Michael A., *et al.* 1995. "Can Standard Cosmological Models Explain the Observed Abell Cluster Bulk Flow?" *The Astrophysical Journal*, vol. 444, 10 May 1995, pp. 507 – 519.

Synge, J. L. 1951. "The Relativity Theory of A. N. Whitehead," *Lecture Series 5. Institute for Fluid Dynamics and Applied Mathematics*, University of Maryland.

---. 1952. "Orbits and Rays in the Gravitational Field of a Finite Sphere according to the Theory of A.N. Whitehead," *Proceedings of the Royal Society of London. Series A, Mathematical and Physical Sciences*, Vol. 211, no. 1106 (Mar. 6, 1952), 303 – 319.

Tanaka, Yutaka. 1985. "Einstein and Whitehead" paper delivered to the Whitehead Relativity Conference held at Claremont twice in November 1984 and in October 1985.
http://www.asahi-net.or.jp/~sn2y-tnk/tanaka_4_0.htm

Taylor, Edwin F. and Wheeler, John Archibald. 1966. *Spacetime Physics*, W. H. Freeman and Company, San Francisco.

Temple, G. 1923. "A Generalisation of Professor Whitehead's Theory of Relativity," 1923 *Proceedings of the Physical Society of London*, vol. 36, pp. 176-193.

Thompson, Sylvanus P. 1987. *Calculus Made Easy*, third edition, St. Martin's Press, NY.

Weyl, Hermann. 1952. *Space Time Matter*, Dover Publications, Inc., NY.

---. 1980. *Symmetry*, Princeton University Press, Princeton.

Whitehead, Alfred North. 1898. *A Treatise on Universal Algebra*, Cambridge University Press, Cambridge.

---. 1906. *The Axioms of Projective Geometry*, Cambridge University Press, Cambridge.

---. "PNK." *An Enquiry into the Principles of Natural Knowledge*, Cambridge University Press, Cambridge. 1919. (Reviewed: G. D. Hicks, *Nature*, vol. 105, June 10 1920, pp. 446 – 448.)

---. "CN." *The Concept of Nature*, Cambridge University Press, Cambridge, (1920) 1971. (Reviewed: G. D. Birkhoff, *Bulletin of the American Mathematical Society*, vol. 28, 1922, pp. 219 – 221.)

---. "R." *The Principle of Relativity with applications to Physical Science*, Cambridge University Press, Cambridge, 1922. (Reviewed: Unsigned, *Nature*, vol 111, May 26, 1923, pp. 697 – 699.)

---. "SMW." *Science in the Modern World*, The Free Press, NY, (1925) 1967.

---. "PR." *Process and Reality (corrected edition)*, The Free Press, NY, (1929) 1978.

---. "AI." *Adventures of Ideas*, The Free Press, NY, (1933) 1961.

---. *The Interpretation of Science (selected essays)*, The Bobbs-Merrill Co., Inc., Indianapolis, 1961.

Wigner, Eugene 1979. *Symmetries and Reflections*, Ox Bow Press, Woodbridge, CT.

Will, Clifford. 1971a. "Theoretical Frameworks for Testing Relativistic Gravity III," *The Astrophysical Journal*, 169: 125 – 140.

---. 1971b. "Relativistic Gravity in the Solar System II. Anisotropy in the Newtonian Gravitational Constant," *The Astrophysical Journal*, 169: 141 – 155.

---. 1973. "Relativistic Gravity in the Solar System III. Experimental disproof of a class of linear theories of gravitation," *The Astrophysical Journal*, 185: 31 – 42.

---. 1993a. *Theory and Experiment in Gravitational Physics*, revised edition, Cambridge University Press, Cambridge.

---. 1993b. *Was Einstein Right?*, Basic Books, NY.

Wolf, Joseph A. 1967. *Spaces of Constant Curvature*, McGraw-Hill Book Company, NY.

Wolverton, Mark. 2004. *The Depths of Space*, Joseph Henry Press, an imprint of The National Acadamies, Washington DC.

Zwiebach, Barton. 2004. *A First Course in String Theory*, Cambridge.

Bei Fragen zur Produktsicherheit wenden Sie sich bitte an:
If you have any questions regarding product safety,
please contact:

Walter de Gruyter GmbH
Genthiner Straße 13
10785 Berlin
productsafety@degruyterbrill.com